Samuel L. Macey teaches English literature at the University of Victoria, B.C., and is Associate Dean of Graduate Studies there. In addition, Dr. Macey is a member of the British Horological Institute, the Institute of Work Study, Organisation, and Methods, and of the council of the International Society for the Study of Time.

John Harrison's no. 1 Marine Timekeeper, 1737, and in center foreground his prize winning no. 4, 1759; on the right Larcum Kendall's copy, K.1, used by Cook and Vancouver; on the left Kendall's K.2, once carried by Bligh in the *Bounty*. (*By courtesy of the National Maritime Museum—on loan from Ministry of Defence—Navy.*)

Clocks
and the
Cosmos

Clocks and the Cosmos

Time in Western Life and Thought

Samuel L. Macey

Archon Books
Hamden, Connecticut
1980

Library of Congress Cataloging in Publication Data

Macey, Samuel L
Clocks and the cosmos.

Includes bibliographical references and index.
1. Horology—History. 2. Clocks and watches in literature. I. Title.
TS542.M23 820′.9′38 79-18891
ISBN 0-208-01773-9

First published in 1980 as an Archon Book,
an imprint of The Shoe String Press, Inc.
Hamden, Connecticut 06514

Printed in the United States of America

For June,
Elizabeth,
and Caroline

Contents

Part IV. *The Reaction of the Romantics*

Illustrations

Preface

Horology is concerned with clocks and watches. This book is about horology and its relationship to literature, in which I include the writings of philosophers, theologians, and poets. Mechanical clocks sufficiently accurate to suit the needs of modern urban man were first produced in England during the third quarter of the seventeenth century. Though the horological revolution of 1660–1760 has had a considerable influence on the language, themes, and forms of literature, this influence has received virtually no attention from critics.[1]

In view of the lack of earlier attention to the subject, I have concentrated on historical survey rather than critical enlightenment. My main purpose is to provide the reader with a broad spectrum of evidence demonstrating the impact of horology on Western and, in particular, on English life—and hence on literature too—during the period that traditional critics have called neoclassical (or Augustan) and Romantic. I hope that this book will make the reader more aware of the influence of horology and will encourage him to seek for further evidence either in the period with which we are here concerned or in our own century, and on the basis of the evidence at his disposal make his own critical judgements, which will not necessarily be in agreement with those that I have tentatively offered. Let me repeat that my main purpose is to provide as wide a historical survey as possible. If by doing this I can transfer to the reader some of my own interest in the subject I shall have succeeded.

My procedure has been to outline the relevant horological developments in the first chapter, and in the second chapter to indicate their remarkable influence on the industrial revolution. From here I proceed, in chapter 3, to some related developments in society and the arts during the Restoration and eighteenth century.

Having set the scene in part 1, I move on to deal with the influence of clocks on philosophers and theologians in part 2. Chapter 4 demonstrates that during the horological revolution a wide spectrum of major Western philosophers, natural philosophers, and theologians (as well as many minor ones) used clock metaphors to explain concepts central to their concerns. Chapter 5 considers the related phenomenon of the rise and fall of the watchmaker God.

In part 3, I move on to the poets, by whom I mean literary artists employing both meter and prose. Chapter 6 is concerned with the widespread use of clock analogies, by poets as well as other writers, for explaining the nature of the universe, of animals, and of man. Chapter 7 deals with the virtual unanimity of poets in attacking the Cartesian concept of clockwork qualities in animals.

During the British horological revolution there were few direct attacks on clocks or clockwork automata in their own right (or, for that matter, against neoclassical rules in literature which prevailed during precisely the same period). The concluding three chapters of the book are concerned with a reaction by artists to clockwork values, after the horological revolution, that parallels the changing emphasis in science from a Newtonian clockwork universe to Darwinian biological evolution. In this spirit, chapter 8 juxtaposes Augustan clockwork with Romantic organicism (after about 1760) in both the language and the form of literature. Chapter 9 deals with the ambivalence of the British novelists, and with the love-hate relationship with technology exhibited by the novelists of the American Renaissance. The last chapter—using examples from such artists as Blake, Baudelaire, Hoffmann, and Poe—demonstrates the remarkable unanimity with which Romantic poets treated both clocks and clockwork automata as though they belonged to the devil.

But if clocks and clockwork automata are characteristics of the Devil, then what may be said both of Western technology and of the horological revolution with which it is so closely related? The technological developments that began in England in the third quarter of the seventeenth century have far more than academic interest. They are of vital concern to every person reading this book who has a watch strapped to his or her wrist, and who has become inextricably bound up with the clock-oriented demands that are an unavoidable concomitant of enjoying the material advantages of Western technology.

It is inevitable that in making a study of this kind one should be indebted to a large number of people. I am grateful to the University of Victoria and the Canada Council for financial aid, and to Corpus Christi College, Cambridge, England for facilitating my research by electing me as a Visiting Fellow in 1972. My first paper on this topic was given to and published by the International Federation for Modern Languages and Literatures in the same year, and I am also grateful to the University of Wisconsin Press (*Studies in Eighteenth Century Culture* vol. 5) for permission to include material on Hogarth. Some of my indebtedness to librarians and the keepers of museums, who have been invariably helpful, is included in the notes. I would also like to mention my gratitude to Professors Henry Summerfield, Patricia Köster, Lionel Adey, and David Park; to Graham Odgers; to James Thorpe III and Helen Tartar who have given me much valued editorial help; to Roger Bishop who has been my mentor in eighteenth-century literature; to Ronald Paulson who encouraged me to relate literature to my earlier experience in commerce and industry; to J. T. Fraser whose unique knowledge of the literature related to time has been placed generously at my disposal; and, finally, to my wife who, as usual, has suffered with remarkable equanimity those inevitable domestic dislocations that accompany the generation and birth of an idea.

PART I

The Historical Background

The Horological Revolution 1660–1760

Introduction

The advances of Western man have been made at the price of an increasing servitude to time. His "progress" in this millennium closely parallels the improvement in accuracy and mechanization of time-measuring devices, the improvement in methods for producing them, and the ever-widening circle of those who were eventually influenced by clocks and watches of their own. To study how, why, and when the major developments occurred is to learn something about our society, our literature, and our thought, and about the Western impact on both the rest of humanity and the human environment. Such a study needs no apology.

Though the history of time measurement is as long as history itself, clocks that were sufficiently accurate and sufficiently plentiful for the needs of urban man in society were the product of the British horological revolution during the century between 1660 and 1760. The pendulum was first used successfully in clocks from 1657, and the balance spring in watches from c. 1674. Within a quarter of a century, the accuracy of clocks improved from an error of some five to fifteen minutes per day to that of as many seconds; the previously unpredictable performance of watches could be regulated to an error of some two or three minutes daily. By the time that Adam Smith wrote *The*

Wealth of Nations in 1776, he could point out that better watches were being produced than a century earlier, and for one-twentieth of the price. Clocks and watches were no longer expensive toys.

The relatively large-scale manufacture which ensued—particularly in the hands of Thomas Tompion (1639?–1713)—involved such industrial innovations as batch production, interchangeable parts, the use of subtrades, and tooth and wheel cutting machines, together with advances in metallurgy and metal working. The English supremacy in the horological revolution of 1660–1760 contributed greatly to the English industrial revolution, usually considered to have begun about 1760.

What man experiences and what man thinks and writes are closely related; so reverberations of the horological revolution are heard throughout the literature of the period. In subsequent chapters we shall be concerned with some of the implications of this revolution. Through the connotations which clockwork and clockwork analogies came to have for society, we shall examine a complex relationship in which the horological revolution and neoclassical literature appear to be more interdependent than has been usually recognized.

First we shall outline the relevant developments in the history of time measurement. In general terms, the history of time measurement is the history of techniques for dividing time with greater and greater accuracy into smaller and smaller divisions. To appreciate more fully the particular impact of horology in the period with which we are concerned, one must begin by taking a brief look at the preceding developments in the direction of greater accuracy.

Though time measurement has become an ever more exact science, its history has not. Dates between A.D. 1000 and the fourteenth century have been suggested for the first mechanical timekeepers. Ward states that the first clock of which we have reliable knowledge is one which was set up in Milan in 1335.[1] The earliest extant examples of weight-driven clocks date from the fourteenth century. The Salisbury Cathedral clock is thought to date from 1386, the Rouen clock from 1389, and the Wells clock from 1392. The Salisbury and Wells clocks have both going and striking mechanisms, while the Wells clock (like the clock at Rouen) also chimes the quarters. Both English clocks can be seen; they are, respectively, in Salisbury Cathedral and the Science Museum, London. Their general similarity and relative technical sophistication would suggest that other clocks had preceded them.

Most of our information regarding early mechanical clocks derives from technical descriptions and literary allusions. Dondi's famous astronomical clock of 1364 was recorded with sufficient accuracy for its complex movement to be reconstructed—there is a fine Dondi reconstruction at the Smithsonian—but the standards of technical description were generally very low.

Literary and other allusions can be even less reliable for dating early mechanical clocks than artifacts, technical descriptions, or manuscript illustrations. The lack of reliability relates to a problem in definition, and would appear to derive from the way in which time measurement developed. When reading old manuscripts, one is frequently alerted only by the context to the meaning of words like *horologium*, *horloge*, or even our *clock* (from *Glocke*). Dr. Johnson, as late as 1755, defines *Horloge* and *Horology*: "Any instrument that tells the hour: as a clock; a watch; an hourglass." He then quotes Brown: "Before the days of Jerome there were *horologies*, that measured the hours not only by drops of water in glasses, called clepsydra, but also by sand in glasses, called clepsammia." [2]

Since there is evidence that before the advent of the mechanical clock there were water clocks fitted with complex wheelwork that even sounded a bell, we cannot be certain that Dante's famous simile in *Paradiso* (early fourteenth century) refers to a mechanical clock: "Then as the horologue, that calleth us, what hour the spouse of God riseth. . . /so did I see the glorious wheel revolve and render voice to voice in harmony. . . ." [3] This early clock analogy—if it is indeed an analogy with a mechanical timekeeper—compares the clock with the movement and music of the spheres. In fact, Price has suggested that the clock developed precisely out of man's need to illustrate the movement of the heavenly spheres;[4] the timekeeping properties were at first only incidental.

In *Heavenly Clockwork*, Needham, Wang Ling, and Price have brilliantly reconstructed the story of Su Sung's astronomical clock-tower of 1088, described (together with illustrations) in 1172. Su Sung's clock-tower was thirty-seven feet high. The time was indicated both aurally and by "jacks" that appeared in apertures in the tower. There was an armillary on the top and a celestial globe inside; both were rotated by the clock mechanism once in a sidereal day. The escapement, which does not concern us here, was a transitional form of water-balance escapement.

The heart of any mechanical clock is its escapement, the device which through a repetitive mechanical motion regulates the running down of the motive power. The history of mechanical clocks is, to a large extent, the history of the improvement in their escapements. The verge and foliot or verge and balance wheel escapement, though not itself isochronous (i.e., vibrating with the uniformity of a pendulum or balance spring) was an invention of genius in its own right.

The verge and foliot escapement was employed in mechanical clocks from their beginnings, until, in 1657, Huygens achieved a major horological advance by substituting the pendulum for the foliot; he used the clock as a mechanical method for both maintaining the movement of the pendulum and counting the number of its relatively isochronous swings.

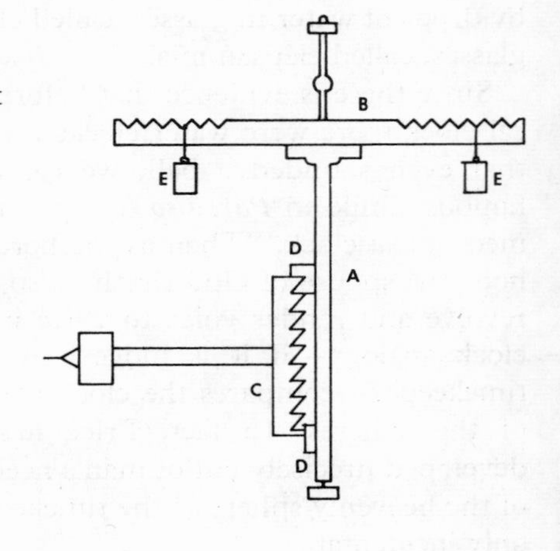

Fig. 1. In the verge(A)-and-foliot(B) or crown-wheel(C) escapement, the weighted foliot bar (frequently replaced by a balance wheel) oscillates back and forth, being actuated by the crown wheel and yet regulating the escape of the crown wheel through the two pallets(D) attached to the verge. If the weights(E) are moved towards the centre, the clock will gain. The motive power for driving the crown wheel can derive from either a weight or a spring (not shown).

The Early History of Time Measurement

Men of the hunting and gathering cultures were more aware of the heavens and the seasons than are men in modern urban society. The needs of harvest and sowing also impose a cyclic pattern on agricultural communities. In recorded history, a calendar, however crude, has been essential to virtually every religion. The apparent annual movement of the sun provided a crude division of a man's life into years and quarter years. The orbital movement of the moon provided a crude division into months and parts of months, as well as having a cyclical effect on water, agriculture, and man. The very names of the days demonstrate the control that the planets were thought to have over men's lives: Saturn's day, the Sun's day, the Moon's day, and—continuing in French—the days of Mars, Mercury, Zeus, and Venus. But the period most readily definable by our ancestors was not the year, the season, or even the month—it was the day. This was the period that really mattered—when one could work or fight, eat or play—the period for which was named the greatest of the gods (Deus, Zeus, Dieu). Though it should be considered within the total context of the calendar, it is with the division of the smaller period that horology is strictly concerned.

The fact that the horological revolution took place in a latitude of over 50°, while the cradle of Western civilization was at least 20° closer to the equator, had some bearing on the course of horology. In northern Egypt, the difference between the longest and the shortest day is no more than four hours. In London, the length of daylight at the time of the summer solstice is considerably more than double its length at Christmas. In Africa, one demonstrates the movement of the sun by describing a vertical arc with one's arm; in London, one describes a horizontal arc. Men measured the division of the daytime by the length of the shadow in countries where the sun was virtually overhead, but in more northerly latitudes they used the direction of the shadow.

Sundials were eventually made with considerable sophistication; but it would seem that as long as they dominated the measurement of time, days (like nights) were divided equally. Except at the equinox, this meant that the hours of the day differed from those of the night, just as the hours of one month differed from those of another. Around the fourteenth century—when public mechanical clocks began to dominate city activity—the earlier "unequal" (or temporal) hours gave way to what were known as "equal" hours. Both terms are to be found, for example, in Chaucer's *Treatise on the Astrolabe* (c. 1391); as late as 1516, More still feels the need to stress that his Utopians use equal hours.

The impact of public mechanical clocks seems to have been the main cause for the adjustment of "sundial time" to "equal time." But the sundial, in many forms, long controlled the nature of time measurement. It had progressed considerably from its beginnings with trees, poles, and perhaps even obelisks. The earliest extant shadow clock is a fragment from Egypt (c. 1500 B.C.) in the Berlin Museum; that museum also has a more complete model in green schist (10th–8th century B.C.). The first sundial seems to have appeared in Rome about 290 B.C.; later writings, like Vitruvius' *De Architectura*, indicate a considerable proliferation of both types and numbers when the empire was at its height. The Saxons, using simpler forms of the sundial, divided the day into four "tides," a practice reflected in such terms as *noontide* and *eventide*.

The history of Japanese time measurement, on which Bedini has done extensive work, offers support for the theory that the equally divided twenty-four hour day is connected with the introduction of the mechanical clock. Shortly after Japan was thrown open to Western influences in 1866, clocks of a modern type entered the country and equal hours were first introduced.[5] A Japanese "Lantern" clock in the Science Museum has two foliots which required adjusting each fortnight in order to allow for the variation in both day and night hours; the changeover from one escapement to the other operates automatically. On other lantern and pillar clocks, the distance between hour numerals had to be adjusted to allow for the variation in the length of the hours. Such complications naturally retarded the manufacture of clocks, despite the fact that appropriate Western technology had long existed.

During the period when unequal hours were used in the Western world, the methods suited to measuring equal hours seem to have been of subsidiary importance. Either the sundial or the nature of the society demanded the use of unequal hours. The Cairo Museum has an early Egyptian water clock (there is a cast in the Science Museum) of c. 1400 B.C. It measures time by slowly losing water from a hole near the bottom. For each month, there is marked a separate series of twelve water levels that divides the night into twelve equal hours. This is a relatively simple system for dividing time that is very much complicated by the need to provide hours of different lengths for day and night according to the changing duration of sunlight. Literary allusions suggest the existence in the past of complex clepsydrae (water clocks) involving wheelwork, jacks, and other automata, as well as the visual and aural marking of time. Clearly these would have been complicated by the need to show unequal hours.

There are also indications that short periods of time may have been measured "equally" even before the advent of equal hours by such methods as the sinking-bowl type of water clock, or the burning of lamps and candles. The story of King Alfred dividing his day by marking off candles is an early example of work study some five centuries before the advent of the mechanical clock. Water clocks, which are known to have been used for limiting lawyers' speeches in classical times, must also have measured equal hours.

One invention of obscure origin may have contributed to the acceptance of equal hours which was so important as a prerequisite for further horological development. Like the mechanical clock, the sandglass is of unknown origin, but all the firm evidence that we have indicates that both began in the fourteenth century. Our first dependable proof for the existence of the sandglass is the figure of Temperance carrying what is very clearly such a timekeeper in a fresco by Ambrosio Lorenzetti (1338). Many of the earliest illustrations of clocks were associated with the cardinal virtue, Temperance.[6] In the more secular society of the horological revolution, we shall find that clocks themselves become the symbols of regularity and order.

Drover cites as the first textual references for sandglasses: "xii orlogiis vitreis" (twelve glass horologes) in 1345–46, and "ung grant orloge de mer, de deux grans fiolles plains de sablon" (a large sea clock with two large phials filled with sand) in 1380.[7]

The simultaneity with the first references to mechanical clocks is remarkable. Since all are called *horloges*, water clocks and sandglasses can only be differentiated from mechanical clocks through context. Unlike the mechanical clock, the sandglass had the initial advantage that both the material and the technology for its relatively cheap production were easily available.

The context of the two textual references given above clearly suggests that they are sandglasses made for taking to sea. Much evidence suggests that sandglasses were first invented in the Mediterranean area for this purpose.[8] Anyone who has sailed by the type of dead reckoning that can be used in the Mediterranean knows that, in addition to the charts (giving the bearings or "winds") and an assessment of the ship's speed, the mariner required a timekeeper unaffected by the motion of the sea. As early as 1306–13, Francesco da Barberino, an Italian poet, says that the careful mariner must have his *arlogio* as well as his chart and lodestone.[9]

Though there is no way of proving that sandglasses contributed to the rationalization of the day into twenty-four equal hours there is much to suggest it. Between Barberino's time and the horological revolution, marine transport went through a revolution of its own. Dead reckoning was no longer enough, for marine activities were increasingly concerned with transoceanic routes. Astronomical navigation demanded more exact timekeeping, and chronometers were the result.

What has been said may suggest that the clock is "nought but a fallen angel from the world of astronomy"; Price maintains much the same in another way. He feels that the first great clocks of medieval Europe were designed as astronomical showpieces. They illustrated the motions of the planets as well as the sun and moon, and carried through the "involved computations of the ecclesiastical calendar." Though they did show the time, this was only incidental. The fact that the early clocks already demonstate a surprising sophistication lends credibility to Price's thesis.[10]

Once equal hours were accepted, a great impediment to the production of mechanical clocks was removed. But, for economic and technological reasons, their wide dissemination came much more slowly than in the case of sandglasses and sundials. The verge and foliot escapement, found in the earliest mechanical clocks, was to remain the standard escapement for some three hundred years; the next important change concerned the motive power. Probably around the middle of the fifteenth century, a spring drive was substituted for weights.[11] Ullyet notes that locksmiths rather than blacksmiths became associated with horology about the same time.[12]

Spring-driven clocks had the advantage of being portable, and this permitted the subsequent development of watches. The earliest watches that have come down to us date from just before the middle of the sixteenth century; in this case, the earliest illustration (1560) is of virtually the same date.[13] It was also at about this time that brass came to be used in watches and portable clocks, instead of iron or the steel that had probably been introduced by locksmiths.[14]

Until the advent of the fusee, the spring drive increased the inaccuracy of clocks. This is because the force transmitted to the gear-train by an uncoiling spring diminishes progressively. The fusee (Huygens calls it a pyramid in the *Horologium*) is a spirally grooved cone-shaped pulley of varying diameter attached to the mainspring by a cord or chain. It provides an excellent method for equalizing the force of the mainspring and is still in use in chronometers. However, since the end of the eighteenth century, problems of size and complexity, coupled with the advent of better escapements, have led to its general demise in other timekeepers. Leonardo illustrates the fusee in a drawing of c. 1490. There is also evidence pointing to a *terminus a quo* between 1455 and 1480.[15]

The Pendulum in Clocks

The heliocentric views of Copernicus, published in *De revolutionibus orbium coelestium* (1543), brought about a revolution in astronomy. As a result, astronomers became ever more insistent in their demands for increased accuracy in time measurement. By the end of the sixteenth century, Jost Burgi, a skilled instrument maker, was attempting to meet the standards required by Tycho Brahe and Kepler. Burgi's crystal clock

(c. 1600)—so called because of its rock-crystal dials—provided separate dials for hours, minutes, and seconds. It is one of the first recorded uses of the second hand.[16]

Meanwhile Galileo was also much concerned with accurately measuring time. The story of the swinging lamps in the cathedral at Pisa is known to most school children. In 1581 or 1582, he is said to have noted the isochronous quality of their swing by timing it against his pulse. (Despite this apparent quality, it should be noted that the smaller the arc of the swing and the less the impulse interferes with the pendulum, the more accurate are its time-measuring properties.) Pendulums maintained by hand were soon being used by astronomers for the purpose of their observations, but it appears to have been many years before Galileo clearly related his idea to clockwork. The clockwork of a pendulum clock can be regarded as an automaton taking the place of the astronomer; the weight or spring-driven mechanism provides the impulse to the pendulum, and also counts the number of its swings. The concept of an automated astronomer may be compared with the tradition of a "jack" striking the bells, that derives from the earliest clocks. In this case, the jack is probably providing an actual illustration of the earlier duties of the keeper of the *Glocke*. Though the automaton is only of peripheral interest to the early part of our study, in chapter 10 we shall have more to say about the Romantic reaction to clockwork automata, which—in such forms as jacks, crowing cocks, or holy figures—have been associated with clocks since their earliest times.[17]

Shortly before his death in 1642, Galileo seems to have conveyed his ideas to his son Vincenzo. Like Brahe and Kepler, Vincenzo worked through a craftsman (Domenico Balestri), but he died in 1649 before the work could be brought to fruition. Galileo's design involved a pin wheel type of escapement that did not come into use until the following century; Huygens—to whom the first successful pendulum clock must be credited—essentially substituted the pendulum for the foliot while retaining the existing but inferior verge mechanism.

Like Galileo, Huygens was directly influenced by his need, as an astronomer, for accurate time measurement. He makes this point in the *Horologium* (1658), and adds: "Astronomers, certainly, are

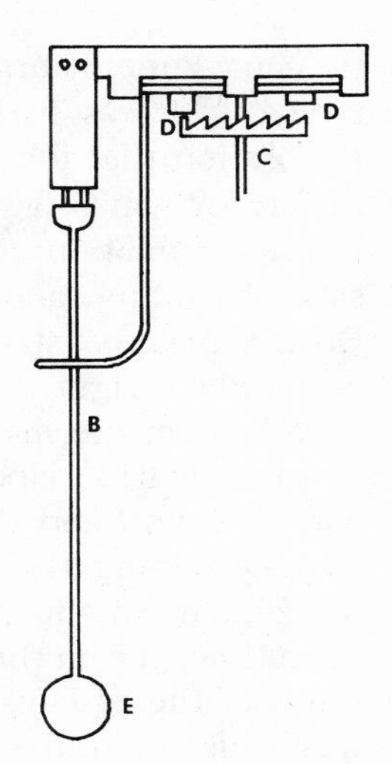

Fig. 2. When Huygens adapted the verge escapement for his 1673 clock, he gave both the verge(A) and the crown wheel(C) a ninety-degree turn (compare with Fig. 1). Through the pallets(D) attached to the verge, the pendulum now regulates the escape of the crown wheel but is also actuated by it. The length of the pendulum controls the time of the swing. (B is the pendulum bar and E is the pendulum weight that can be adjusted.) Theoretically a pendulum of 39.14 inches will have a one second swing in London.

adopting it [the pendulum clock], so that henceforth there will be no troublesome urging of pendulums nor watchful counting required."[18]

The Spring Balance in Chronometers, Watches, and Portable Clocks

The relationship between horology and modern civilization has been frequently noted, though more detailed research is warranted. Lewis Mumford states: "The clock, not the steam engine, is the key machine of the modern industrial age." Commander Waters puts the case for the importance of navigation: "It is time which makes modern civilization practicable. But it is the provision of accurate time in ships at sea which lies at the core of civilization for its wealth is largely dependent upon the safe and timely passage of ships at sea."[19]

Bruton makes an even more specific claim concerning the prize of twenty thousand pounds offered in 1714 for the discovery of a method to ascertain the longitude at sea: "The act of 1714 caused the same kind of surge of scientific effort that space research does today, and was in many ways responsible for the Industrial Revolution that followed. The invention of the marine chronometer, for which it was directly responsible, resulted eventually in the domination of the world by the British Fleet, the expansion of trading, and the acquisition of the British Empire."[20] Bruton may be stating his case a little too strongly—but in essence he is right.

In layman's terms, the reason why it was crucial to invent an accurate marine clock for discovering longitude is as follows: with a sextant and a compass, a mariner can obtain his latitude by ascertaining the angle between the sun, the ship, and the horizon, when the sun is directly to the north or south (local noon time). From this he knows how many degrees he is north or south of the equator (his latitude). To find his position east or west (his longitude) is, however, a much more difficult task because the earth, in spinning on its axis, upsets astronomical calculations. But the earth spins at a relatively even rate through 360° in a period that we have divided into twenty-four hours. If therefore a mariner had with him an accurate clock regulated to the time at Greenwich, England, and found that (in the Atlantic) his clock indicated 1:00 P.M. when the sun was directly to the north (local noon time), he would know that he was somewhere on the longitude 15° (360° divided by the difference in time) west of Greenwich. Since he could also ascertain the latitude, he would thereby know his precise position.

As early as 1530, the Flemish scientist Gemma Frisius had recommended the use of a watch at sea for determining longitude, but until the advent of the pendulum no sufficiently accurate mechanical instrument existed. That this matter was never very far from the minds of scientists may be concluded from the closing request in Newton's letter to Aston, as early as 18 May 1669. Newton wished to know from the Dutch whether "pendulum clocks doe any service in finding out ye longitude &c."[21] Sprat, in his *History of the Royal Society* published two years earlier, says optimistically, "There is only wanting the *Invention of Longitude*, which cannot now be far off."[22] Huygens was fully aware of the problem. At the beginning of the *Horologium*, he says: "The so-called science of longitude, which,

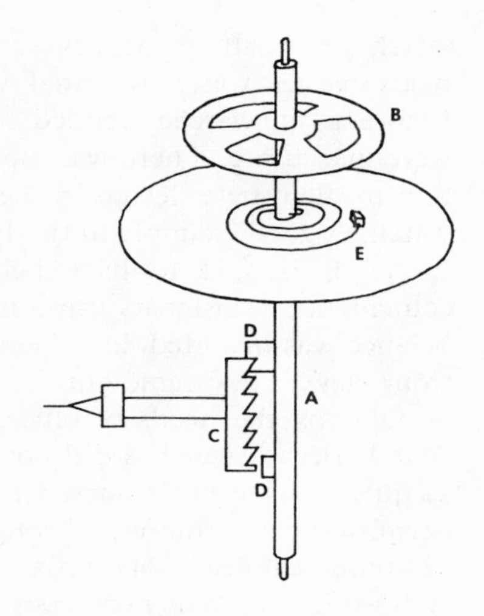

Fig. 3. For his balance-spring regulator, Huygens adapted the verge or crown-wheel escapement in a manner comparable to that which he had used with the pendulum. (The balance wheel had sometimes been substituted for the foliot bar in earlier verge escapements.) By attaching a spring(E) to the verge(A) Huygens achieved both portability and near accuracy. In a manner comparable to the previous two (schematic) diagrams of the verge escapement, the verge and its pallets(D) alternately engage successive teeth of the crown wheel(C) as the balance(B) oscillates back and forth.

if ever it existed, and so had provided the greatly desired help to navigation, could have been obtained in no other way, as many agree with me, than by taking to sea the most exquisitely constructed timepieces free from all error. But this matter will occupy me or others later...."[23]

It soon became clear that chronometers, like watches, could not operate satisfactorily with pendulums. Through trying to find a suitable escapement for chronometers in seagoing conditions, Huygens invented the spring balance in 1674–75.[24] In essence, Huygens was substituting the isochronous qualities of the spring for those of the pendulum. Much as he had done with the pendulum, Huygens adapted a new concept to existing technology. The balance wheel (without isochronous qualities) had long been an alternative to the foliot.

Much as the pendulum gave accuracy to clocks, the spring balance gave both portability and accuracy to the earlier portable clocks and to watches. The portability of the early spring-wound table clock had made it the natural ancestor of smaller but still cumbersome watches. One can trace this development from the drum-shaped table clock through the smaller carriage clock of similar shape (used in carriages) to the thick round carriage watch and finally to pocket watches of comparable shape.[25] The fusee that the watch inherited was one of the reasons for its bulk. Before 1675, watches tended to be as highly decorated as they were inaccurate. There was, however, a somewhat earlier exception to the ornate design in the form of a plain silver "Puritan" watch. A good example in the Fitzwilliam Museum (Cambridge, England) is said to have belonged to Oliver Cromwell. By coincidence, waistcoats came into fashion just when the spring balance was invented, and thereafter men's watches disappeared from view. They came out again when wristwatches were made popular by the needs of officers in the trenches during World War I. British watches did not lose their cumbersome thickness as quickly as French ones. In France, the cylinder escapement permitted extra thinness through the abolition of the fusee; the "Lepine calibre" of 1765 and the watches of Breguet (1780–1820) became progressively slimmer.

Some of the desire for luxury that was satisfied by ornateness in the earlier less accurate watches found a new outlet in the pride of possessing repeater watches. Repeater clocks and more particularly repeater watches enthralled many members of eighteenth-century society. By communicating the time at the press of a button, they derived much of their fascination from qualities comparable to those of automata. Repeater watches were made to indicate aurally the nearest hour, quarter, half-quarter, five minutes, or even minute. Though the period of development was much shorter, the history of repeaters parallels that of mechanical timekeepers as a whole. The repeater movement began with clocks and then was adapted to watches; there was also a chronological development in the direction of indicating progressively smaller divisions of time. Repeating work for clocks was invented about 1676; in about 1687, Quare made a repeating mechanism for watches, as did Tompion, working to the design of the Rev. Edward Barlow, who is credited with the repeater clock. There followed a succession of increasingly sophisticated repeaters until Mudge's repeating mechanism of 1757 (believed to be the first of its kind), which could signal the precise minute.[26]

Though there have been several unsuccessful attempts to mass produce repeaters, they have always remained a luxury article from which their owners derived a particular pride. The crude ancestors of the repeating watch were the mechanical timekeepers whose hands and numbers could be felt in the dark. One of the earliest monastic clocks (c. 1390) is marked with sixteen Roman numerals; the typical single hour hand permits the approximate hour to be "read off" during the night by touch.[27] The scarcity of artificial light made repeaters the great social wonder of our first mechanical age. Safety matches from the middle of the nineteenth century (Lundström's patent of 1852) and luminous watch dials after the technical advances of World War I have combined with domestic electric light to terminate the production of repeating watches. They are one of the technological wonders that have fallen a prey to subsequent advances in technology.

We have observed thus far how, on serveral levels, time measurement has involved the division of time into progressively smaller units as timekeepers became more and more accurate.[28] One can also note that beginning with the horological revolution domestic time measurement involved the production of small mechanisms in ever increasing quantities. We shall be concerned with this development in the next chapter where we deal with the relationship between the horological revolution and the subsequent industrial revolution.

The Horological Revolution and the Industrial Revolution

The Production of Clocks and Watches

Huygens assigned his rights in the pendulum clock to the tradesman Salomon Coster, who took out a patent on 16 June 1657. What occurred after this event shows how ready the tempo of technology was for increased acceleration. In September 1657 John, the eldest son of the clockmaker Ahasuerus Fromanteel, went from England to Holland to work with Salomon Coster. The first known English pendulum clock is signed on the backplate, "A. Fromanteel, London Fecit 1658."[1] What is noteworthy is not so much the rapid spread of ideas as the rapid incorporation of such ideas into practical production.

Until this time, clocks and watches were expensive and decorative toys. The pendulum clock provided the first reasonably accurate method of mechanical timekeeping and this coincided with accelerated urbanization. Before 1657 clocks could not generally keep time more closely than to about fifteen minutes per day; within twenty years they could frequently be relied on to vary by less than ten seconds per day.[2] The revolutionary factor is that for the first time in man's history it was possible to produce timekeepers accurate enough for any normal domestic purposes of urbanized man. Intellectually, the point of change lies at least as far back as Galileo, but in the domestic and urban sphere the advent of the pendulum clock is decisive.

The modern mechanical age had arrived, and the techniques as well as the thoughts of men would never be quite the same again. In his article, "The First Twelve Years of the English Pendulum Clock," Michael Hurst makes the statement, supported by examples: "One of those at present unexplained facets of early clockmaking is the incredible similarity between pieces which bear different names, so much so that one cannot but have the feeling they were made at the same bench."[3] Not only clock cases but also designs and parts were beginning to show similarities. Each clock was still, in some measure, an individual artifact overcoming individual problems, but the ever increasing repetition imposed by demand compelled craftsmen to rationalize both design and production.

There was an element of feedback involved. In a situation of increasing potential demand, the rationalization of production methods reduces prices. This increases demand, and produces pressure for further rationalization. Throughout the eighteenth century, the production of watches by the method of division of labor was pressed as far as it would go. Despite the fact that the cost of living had at least doubled during that period, Adam Smith, in 1776, was able to use watches as the most impressive example of what such production methods could do. We have earlier alluded to his point that a better watch than that which had cost twenty pounds about a century earlier might in the latter half of the eighteenth century be purchased for twenty shillings. Smith notes that this has astonished "the workmen of every other part of Europe," and he adds that "in the clothing manufacture there has, during the same period, been no such sensible reduction of price."[4] During the industrial revolution, the textile trades could and did learn much from the advances in horology.

Smith also demonstrates by implication that a great deal of the glamour was going out of watchmaking. One of his arguments against restrictive practices is well known: "People of the same trade seldom meet together, even for merriment and diversion, but the coversation ends in conspiracy against the public, or in some contrivance to raise prices." Less well known is his argument against apprentices taking seven years to learn what "cannot well require more than the lessons of a few weeks." He uses the examples of clocks and watches to demonstrate that though the original inventions were "the happiest efforts of human ingenuity," repetition and the division of labor have resulted in the trade containing "no such mystery as to require a long course of instruction."[5]

There are few reliable statistics about watch production during the eighteenth century. Wright's statement that "In 1703 Clerkenwell alone made 50,000 watches for home sale and 120,000 for export" seems remarkably high for that period, and is not supported by documentation.[6] It has been estimated with some reliability that Tompion (1639?–1713) made approximately 6,000 watches and 550 clocks during his lifetime.[7] He is credited with the process of batch production—involving relative interchangeability of parts—that represented a considerable advance in producing mechanical artifacts without any necessary reduction of standards. Symonds says that: "In order to achieve this large output, Tompion organized this workshop in a way hitherto unknown in the English handicrafts." Symonds (the biographer of Tompion) quotes what Sir William Petty, the greatest political economist of the seventeenth century, had to say about the advantages which would accrue from a division of labor: "'In the making of a Watch,' wrote Petty, probably with *The Dial and Three Crowns* [Tompion's workshop] in mind, 'If one Man shall make the *Wheels*, another the *Spring*, another shall engrave the *Dial-plate*, and another shall make the *Cases*, then the *Watch* will be better and cheaper, than if the whole Work be put upon any one Man.'"[8]

Charles Babbage—the Lucasian Professor of Mathematics at Cambridge from 1828 to 1839, and originator of many of the ideas underlying the modern automatic computer—points to a continuing high output in 1798, despite the durability of watches: 50,000 per annum were being produced for the home market in addition to the very important export trade.[9] West estimates British production in the early nineteenth century at 120,000 clocks and watches worth six hundred thousand pounds per annum and employing twenty thousand persons.[10] In 1832, Babbage still uses watchmaking as the most imposing example of the division of labor. What he says helps to explain Adam Smith's surprising contention that watchmaking was not a difficult trade to learn. According to Babbage there were in his time a hundred and two distinct branches of the watchmaking trade, to each of which a boy might be put apprentice. The division of labor had become so specialized that the watch-finisher, who assembled the parts, was apparently the only tradesman who could work in any other department than his own (pp. 162–63).

But within a generation, the British watchmaking industry was doomed. This was essentially because, having a vested interest in established skills and methods, it was not prepared to take the next leap forward. Specialization had gone so for that the individual specialist would have been merely confused and slowed down by the introduction of new mechanisms. What was needed (as has occurred more recently with the Model T's of Ford or the Liberty Ships of Kaiser) was a completely new look at the trade by those who could adapt the simplified procedures of the division of labor to the next step in achieving a fully mechanized production.

Applying lessons learned from the manufacture of small arms, American watchmakers, followed by Swiss and French ones, introduced mass-production methods. (*The American Clock 1725–1865* offers an excellent pictorial introduction to early American clocks.)[11] In America, the watchmaker Aaron L. Dennison joined forces with Eli Whitney, who made rifles by machine at his Springfield plant. Their company, eventually called the Waltham Watch Company, fathered most of the other American factories. The mechanization of the watch industry produced even cheaper watches despite very considerable inflation. It took Dondi sixteen years to create one clock; Tompion made about six thousand watches by batch production in his lifetime; a century later, England was producing over one hundred thousand watches per annum by division-of-labor methods; but today Switzerland alone is exporting about fifty million watches per annum. In addition, she is now being rivalled in the low price market by Russia and Japan as well as America. Indeed, North America currently consumes more than forty million watches per annum, many of them cheap "fashion" lines.

Despite the fact that these figures are approximations, they very clearly demonstrate a remarkable growth in the production of a relatively durable artifact. Dondi's clock was priceless, but Tompion lived at the beginning of the modern age—he had a price list and charged twenty-three pounds for an ordinary watch in a gold case, eleven pounds for one in a silver case, and seventy pounds for a gold repeating watch.[12] This was at a time when servants could be bonded for three pounds per annum, and many a curate was expected to support his wife and family on twenty pounds per annum. Despite an inflation of perhaps fifty-fold that has since taken place, the price of a watch is no greater today than it would have been three hundred years ago. Moreover, the annual consumption of watches is a measure of the incredible spread of the franchise that has taken place. There is an undoubted relationship between technological progress—of which the watch continues to be an essential element—and the ever widening spread of the franchise to all nations, colors, and creeds. Yet a price has to be paid. Not only production, but also men become slowly though inevitably mechanized.

The Horological and the Industrial Revolutions

There has been some but perhaps not enough recognition of the importance to Western technology of Protestants in general, and Puritans in particular.[13] The traditional relationship between Temperance and the clock clearly carries over into Protestant ethics. Also, one is impressed by the number of clockmakers who are Dissenters. But, as with key scientists nowdays, numbers are not the only criterion. Huygens himself, being a Protestant, left France as a result of conditions connected with the revocation of the Edict of Nantes in 1685. At that time, many Huguenot watchmakers emigrated to Switzerland, Holland, and England. Their influence on the horological revolution and on technology would be well worth further research. So, too would the influence of the horological revolution on the industrial revolution, after 1760. Specific instances come readily to mind. We have already noted the contribution made by watchmaking in adding batch production and relative interchangeability of parts to the type of division of labor earlier involved with comparatively simple articles like needles and pins.

A further contribution made by clockmaking is related to the published descriptions of its processes. Some of the earliest works on modern technology deal with clockmaking. The ability to illustrate and describe industrial processes clearly and accurately is of vital importance for technological progress. (We shall see later that it is by no means unconnected with the clarity and order that became an ideal for language at much the same time.) Huygens' *Horologium* (1658) and the much greater *Horologium oscillatorium* (1673), written in Latin, deal essentially with specific improvements to horology; William Derham's *Artificial Clockmaker* (1696) deals with horology's history and processes, is written in English, and is addressed to artisans as well as virtuosi. In the same period, there were a number of lesser publications, such as John Smith's *Horological Dialogues* (1675).

Diderot's *Encyclopédie*, published at the beginning of the English industrial revolution, provides the most impressive proof of the importance of horology among industrial processes as a whole. Diderot is proud to announce that he is offering the stupendous total of over six hundred folio sheets of industrial illustrations of a high order, compared with thirty in Ephraim Chambers' English encyclopedia.[14] Diderot includes sixty-four pages of diagrams dealing with horology and its related machine and hand tools. The *Encyclopédie* divides clockmaking into sixteen and watchmaking into twenty-one proceeses, each of which is described clearly and in detail. The point is then made that this division in the producing of horological parts means that a good master-watchmaker need only study the principles of his art, control his workmen, and overlook their work. In addition, each part of a clock or watch must be perfect because it is made by someone who does nothing else.[15]

This type of analysis is the first step towards industrial engineering and the time and method study which it involves.[16] Indeed the studies on pins by Peronnet in 1760 and Babbage in 1832 carefully analyze the cost of production in terms of the material and labour involved in each stage of the process.[17] The step from careful analysis to suggesting methods for improvement is a logical and a relatively small one; the authors of the

Encyclopédie are conscious of the importance of their work to the development of industry.

The general interest in the mechanical arts was very considerable; the *Britannica* itself began in 1768 as a "dictionary of the Arts and Sciences." Such an interest is part of a continuing movement for upgrading the manual trades first urged by Bacon, the father of modern science, and subsequently by such of his followers as Webster, Sprat, and Boyle. We have earlier referred to the relationship between research and production, giving as examples the astronomers and the craftsmen to whom they transmitted their ideas. Although there has been much controversy regarding Hooke's claims to certain horological inventions, the relationship between the irascible curator of experiments at the Royal Society and Tompion, the father of English clockmaking, is perhaps particularly symbolic of this movement.[18]

To the extent that the industrial revolution implies the application of power machinery to all or most of the manufacture of an article, the British horological revolution can only be a precursor to what occurred from about 1760. But the horological revolution involved the invention and improvement of hand-driven machinery for cutting such watch parts as screws, fusees, and wheels. This increased both the production and the interchangeability of parts.

Another approach to the influence of the horological revolution would be to consider some of the inventions essential to modern technology that were a by-product of its research. Metallurgy has gained in many ways from the work done with respect to compensating for temperature variation that led to Graham's mercury and Harrison's grid-iron pendulums. In fact, the bi-metallic strip that Harrison used for this purpose in his third chronometer has developed into the ubiquitous thermostat of modern society. Other "modern" devices which began with eighteenth-century horology are the feedback of Breguet's *pendules sympathiques*, and the self-winding watch made by Breguet from about 1780.[19] Even the differential gear must be directly credited to horology.[20] Timing and, indeed, automatic timing devices were not unexpectedly influenced by the horological inventions of that period. In this connection, it is worth noting that the earliest factory time clock should not be credited to our own century. It was invented in 1750 at the beginning of the industrial revolution by John Whitehurst of Derby. The stop-watch (used in factory time studies before the end of the eighteenth century)[21] has its origins even earlier in the "pulse watch" (c.1690) of Samuel Watson, intended, as the name suggests, for the use of physicians.[22] (Galileo's pulsilogium was a distant ancestor in function if not in form.)

The horological revolution, as Daumas tells us in *Scientific Instruments of the 17th and 18th Centuries*, "also had very important repercussions on the instrument-making industry.... precision mechanics is in [clockmakers'] debt for its first successful constructions and its basic progress until the present time. For five centuries clockmakers in every generation were responsible for the most exact mechanisms known.... Thus, quite apart from specialized inventions, thanks to which clocks and watches continually increased in precision, the clockmakers were also responsible for putting at the service of mechanics greatly improved tools which the instrument makers could adopt."[23] Graham (one of the fathers of horology) was as famous for his astronomical instruments as for his clocks and watches.[24] But though scientific instruments were important, horology is our main concern; it affected people and poets in ways that scientific instruments could not.

One of the difficulties in assessing the full impact of clockmaking was that, despite the evolution of major centers in London, Coventry, and Prescot, its operatives were very widely dispersed.[25] Moreover, watchmakers have provided an important reserve of precision engineers for other trades. (The British government's assistance to Newmark and Smith and Sons during World War II was influenced by this factor.) Several minor trades, such as the making of gas meters in Clerkenwell, are directly connected with the previous existence of watchmakers in an area. Sheffield, the English Pittsburgh, owes much of its prosperity to the method of casting steel discovered by the clockmaker Benjamin Huntsman, who settled near there in 1740. Huntsman is said to have been led to his discovery by the difficulty of obtaining finely tempered steel for watch springs.[26]

But the textile rather than the steel industry is considered to be the most important pointer to the British industrial revolution. The key inventions—Hargreave's spinning jenny and Arkwright's water frame—both came into use in 1768. Arkwright patented the water frame in 1770 and a number of minor inventions in 1775, but at a trial in 1785 judgement was given against these patents. It turned out that Highs and Arkwright both claimed to have invented water frames, which had almost certainly been made by the Warrington clockmaker, John Kay. It is worth noting that the mechanism of Arkwright's frame was known as "clockwork," and that in his first patent he described himself falsely as "Richard Arkwright, ... clockmaker."[27]

Babbage's story about the boom in machine-made lace is particularly revealing. It centered in Nottingham in the twenty years before 1830, and employed "above two hundred thousand." In making the new machinery, "those who were best paid, were generally clock and watch makers, from all the district round."[28] Much evidence suggests how important, perhaps even how crucially important a large indigenous supply of clockmakers available for the manufacture and maintenance of machinery was for the British industrial revolution.[29]

The Rationalization of Time and Timepieces

The results of Huygens' pendulum and spring-balance escapements were revolutionary. But the process was part of an evolution (albeit an exponentially progressive one) to which Western technology seems irreversibly attached. The first pendulum escapement improved the accuracy of clocks from between five and fifteen minutes per day to within ten seconds per day. By 1761, Harrison's fourth chronometer erred by no more than fifteen seconds after a five-month journey to the West Indies and back.

We have noted how man's search for an exact time measurement has progressed through the years, seasons, months, and days of the calendar to even smaller divisions of time. Minutes, and later seconds, did not normally appear on clocks until after the discoveries of Huygens. By early in the twentieth century,

Shortt's free pendulum clock had virtually exhausted the potential accuracy of mechanical timekeeping. The quartz crystal clock, and, even more recently, the caesium atomic clock and hydrogen maser, provide standards of accuracy required by interplanetary travel and stellar astronomy rather than urban society.

Increasing accuracy has demanded the rationalization of time itself. Just as mechanical clocks demanded a division of the day into equal hours, so the accuracy of the pendulum clock led to the general adoption of the mean rather than the solar day. Ordinary people had become aware that the apparent solar time of the sundial could vary by as much as sixteen minutes from the mean time provided by clocks. In the nineteenth century, the speed of railway transport forced a further rationalization of time measurement. Sandford Fleming in Canada and Charles F. Dowd in the United States were the principal proponents for grouping local times into time zones. The simultaneous distribution of time measurement, which has been one of the main horological functions of electricity, was closely connected with the railways and their time-tables. In our own day, the accuracy of time measurement has reached a variation of no more than one second in thirty thousand years. In 1967, the second itself was redefined in terms of the vibration of the caesium 133 atom. Astronomers and horologists have discovered that the movements of the planets are not sufficiently accurate for some of the requirements of modern science.

Watch manufacture itself has also been the source of rationalization in industry. Watch manufacture has been essential to modern production methods and the rationalization that this involves. There is an individualism about early watches that both derives from and imparts some of the excitement of overcoming individual problems. The individualism, however, slowly disappears into the monotony of imitation and repetition. Repetition—the essential prerequisite for cheaper production—is related to the increasing demand for relatively accurate domestic time measurement. An aspect of eighteenth-century watch production that should not be underestimated is that sandglasses and pocket sundials—and to a lesser extent clepsydrae, and public sundials—continued to be made in very large numbers because of their relative cheapness. Their owners and users provided the potential customers that permitted the production of mechanical timekeepers in ever increasing numbers.

Unlike the Americans, who were to enter first clockmaking and then watchmaking with a vengeance during the nineteenth century, England seems to have been unable to adapt herself to the new methods of production in horology. The nature of technological progress is such that those who are first in the race can ultimately suffer as a result. In developments related to the industrial revolution, Britain enjoyed early starts in road making, the cotton industry, and railways. This appears to have made change more difficult for her than for other countries in the twentieth century. During the horological revolution, the world thought of England as it more recently thought of America, and now thinks of the dynamism of Japan. The speed with which England entered into the production of watches and clocks at the beginning of the horological revolution, was impressive by the standards of the time. But the method of division of labor may, to some extent, have held back advances like dispensing with the fusee and developing further Mudge's lever escapement (c.1755), the real production of which was to stagnate until the following century.

The subdivision into various trades, as well as the building up of large stocks of watch parts, was bound to inhibit changes in production. During the second half of the nineteenth century, the vested interest of the British watch industry in a high quality hand-finished product virtually priced the British watch trade out of the world market. The British Horological Institute was founded in 1858, in great measure to alleviate the distress of British watchmakers. This occurred almost exactly two hundred years after Huygens' first pendulum clock, and the beginning of England's golden age of horology.

The dramatic rise and fall in Britain's share of the watch industry between 1660 and 1860 with which we have thus far been mainly concerned is the backdrop to much of the following study that deals with the influence of horology on literature. We have also noted the surge of productivity in the New World towards the end of the period. This may help to explain why writers of the American Renaissance, like Whitman and the younger Emerson, were sometimes less completely negative in using the image of clockwork than were the English Romantics of the early nineteenth century.

In this chapter we have been concerned with the relationship between the horological revolution and industry. But poetry, like all art, is in some measure a reflection of life (which means the social as well as the industrial life of a civilization), and this was never more true than in the eighteenth century. Before, therefore, we turn to look at the influence of horology on philosophers, theologians, and poets, we shall consider, in the next chapter, some aspects of the way in which the British horological revolution affected society, and in particular the life of London. Since the topic is a large one, I shall restrict myself to a general overview of the pervasive interest in timepieces during the horological revolution, followed by related studies of three famous Londoners: the father of chemistry, Robert Boyle; the tradesman-novelist Daniel Defoe; and the artist William Hogarth.

The Horological Revolution and Society

The Pervasive Interest in Timepieces

The horological revolution influenced most levels of urban society from kings on down. For a period that is brief by historical standards, a relatively modest trade enjoyed a level of publicity something akin to that which has been achieved by space technicians in our own time.

In the eighteenth century, the leading men of the age were proud to be associated with watchmaking. The king of Prussia, the king of Poland, and even the Emperor K'ang Hsi of China set up clock factories. Leibniz[1] and Benjamin Franklin, who gave us the phrase "Time is money," were, like Descartes, actively involved with clock mechanisms; Louis XVI and Voltaire invested heavily in watch factories;[2] and it was not by writing *The Marriage of Figaro* or *The Barber of Seville* that Beaumarchais raised a fleet of forty vessels to help the American War of Independence—the money came largely from his profits as a brilliant watchmaker.[3] The kings of England, too, had a personal interest in horology. One might adduce many incidents, but I shall restrict myself to two: James II arbitrated in person when the clockmaker Daniel Quare successfully opposed the claim by Edward Barlow to patent his repeater watch (made by Tompion), and George III intervened with the Admiralty in favour of John Harrison after the former had been remarkably tardy in completing payment of the prize of twenty thousand pounds long overdue for Harrison's invention of the chronometer. Tompion and his former apprentice Graham—whose work almost spans the horological revolution—are frequently termed the fathers of English clockmaking. Both tradesmen are buried in Westminster Abbey.

The diarists make frequent allusion to clocks. For Pepys and Evelyn, they were objects evoking personal pride and mechanical curiosity. In general, they seemed worthy of note when this would not always be the case today.[4] Pepys—who was himself later to be the president of the Royal Society—remarks on a visit to an earlier president, "I to my Lord Brunker's, and there spent the evening by my desire in seeing his Lordship open to pieces and make up again his watch, thereby being taught what I never knew before; and it is a thing very well worth my having seen, and am mightily pleased and satisfied with it."[5]

As one might expect, Sprat's *History of the Royal Society* (1667) frequently lists horological inventions.[6] And Sprat asks the rhetorical question, "In what Subject had the wit of *Artificers* bin more shewn, than in the variety of *Clocks* and *Watches*?" Prophetically, he sees "this *Mechanic Genius*" as a panacea for all mankind—even Asians and Africans.[7] A century later, Keysler, in *Travels* typical of the voluminous output of such books, reports on "automata" and "remarkable" or "curious" clocks in town after town. Travellers reported on time measurement much as today they might discuss currencies: ". . . at first it is a little puzzling to reconcile the *Italian* clocks with the *French* and *German* method of computing time, . . ."[8]

Robert Boyle: The Widespread Use of Horological Metaphors by the Father of Chemistry

The professional relationship of philosophers and scientists (or natural philosophers) with horology is of particular importance, and, in the age of mechanistic philosophy and the Watchmaker God, is a subject in its own right, with which we shall be concerned in part 2. But such men—as the philosopher in Molière's *Bourgeois gentilhomme* was currently demonstrating—were also human beings. The fascination that clocks had for Boyle, the father of chemistry, is a case in point, and is reflected throughout his works. Boyle so frequently employed clock metaphors in his explanations that he became self-conscious about their use. In a manuscript at the Royal Society, Boyle introduces a clock metaphor by saying: "To explain this a little, let us resume the often mentioned, and often to be mention'd Instance of a Clocke," A little later, he refers to the Strasbourg clock "I have so often alluded to."[9]

As far as possible, Boyle attempted to avoid hypotheses; he desired rather "to collect Experiments for more rational and philosophical Heads to explicate and make use of."[10] His first published work reports in the new clarity of style that was to be recommended by the Royal Society. The *New Experiments* (1660) deals with the "pneumatic engine," or air pump, completed in 1659 with the assistance of Hooke. Both men were much involved with horology; this is how Boyle opened the lengthy report on his experiment: "We took a watch, whose case we opened, that the contained air might have free egress into that of the receiver. And this watch was suspended in the cavity of the vessel only by a pack-thread." As the air was removed, the sounds of the watch grew fainter and fainter, "though we could easily perceive, that by the moving of the hand, which marked the second minutes [the earlier term for *seconds*] and by that of the balance, that the watch neither stood still, nor remarkably varied from its wonted motion. . . ."[11]

The great advantage of Boyle's prose is that we can visualize his experiment. We have a much less precise image of two Spanish water-powered engines that Digby attempted to describe less than twenty years earlier in his *Two Treatises*.[12] (The ability to document carefully what one sees is of far more importance to science, technology, and art than a "creative" age is generally prepared to recognize. We shall discuss in chapter 8 the influence that science and horology had on language itself at this period.) Like the good experimenter that he was, Boyle returned to his subject again and again, attempting to eliminate the possibility of error by changing the method of approach.

He returns to the watch experiment once more at a later date, this time "suspending in the receiver a watch with a good alarum. . . ."[13]

Boyle's *New Experiments* were concerned with giving animals the same treatment as watches. On one occasion, the bird involved would surely have proved that its mechanism was more delicate than that of a watch had it not been for the "pity of some fair ladies . . . who made me hastily let in some air at the stop cock. . . ."[14] Though the Cartesian theory of the beast as a clockwork machine proved a useful rationalization for animal dissection,[15] Boyles's studies in physiology were restricted by the "tenderness of his nature."

Boyle's experiments with the air pump resulted in an important discovery. It was while answering the objections of Franciscus Linus, a Jesuit critic of the *New Experiments*, that the chemist formulated Boyles's Law: the product of the pressure and the volume of a gas at constant temperature is constant. But Boyle seems to have also been interested in watches for their own sake. He writes of feeling a striking watch "through the several linings of my breeches" and explains at length a new method "to give small glasses the shape that is requisite to fit them to serve for covers to the dial plates of watches. . . ." Boyle was intrigued with the idea that watches might be permanently perfumed. Elsewhere, he recalls a relevant anecdote of a famous physician "who was skilled in perfumes. . . ."[16] Even when he is not dealing with watches himself, Boyle is glad to hear of the latest developments.[17]

In his plea for the value of the newly developing cooperation between science and the "mechanical arts," Boyle holds back for his climax the great advantage that watchmakers derived from the invention of the pendulum clock: "We daily see the shops of clockmakers and watchmakers more and more furnished with those useful instruments, pendulum-clocks, as they are now called, which but very few years ago, were brought into request by that most ingenious gentleman [Huygens], who discovered the new planet about Saturn."[18]

Even Boyle's most solemn moments can be concerned with watches. His *Meditations* offer considerable insight regarding the man who dominated English science in the latter half of the seventeenth century; they were influential enough to invite the burlesque of Swift's *Meditation upon a Broomstick* many years later. In "Meditation VIII: Upon Telling the Strokes of an Ill Going Clock in the Night," Boyle recounts his experience with a watch and a clock, and relates this to utilitarian purposes. In "Meditation IX: Upon Comparing the Clock and His Watch," the same experience provides the material for more philosophical considerations culminating in a man's relationship with his Maker and his eternal life.[19]

Men have always been interested in time. But in the period with which we are concerned a wider range of people than ever before had a particular interest in the mechanical clock. The horological inventions of the architect, Christopher Wren; the writing of the first significant manual of horology by the clergyman, William Derham; and the extensive horological references in the writings of the antiquarian William Stukeley are typical of this widespread interest, as well as of a movement that marks the beginning of our modern relationship between research and industry.[20]

Daniel Defoe: The New Social Influence of Watches

Watches, as artifacts, were coming to have greater meaning for the larger and more materialistic urban civilization developing in London. We have already noted that a watch represented as much as ten years' income to a servant girl. Clearly they were an article that could evoke both pride and desire. Some of Defoe's novels of the 1720s offer as good a picture as any of the seamier side of London life. This has been a subject for literature at least since Greene's tracts on "Conny-catching" (1591 and 1592), and even the "Canon Yeoman's Tale" of Chaucer. What has changed is the importance of the watch as a substitute for ready money.

Before turning to Defoe, let us look briefly at Farquhar's *Inconstant* (1702) and Goldsmith's *Citizen of the World* (1762) in order to demonstrate that the activity of "taking a watch" is by no means limited to a single decade. In the last act of Farquhar's play, Lamorce, "A Woman of Contrivance," produces four

"Bravo's" who are to relieve the old fool Mirabel of his valuables. When she asks him what time it is, he tries to pretend that he has left his watch at home, but his memory is jogged by one of his men. "O dear Sir, an English Watch! *Tompions* I presume," Lamorce says as the watch is produced. But this is a comedy; in the typical reversal of 5.4, after the "Bravo's" have been outnumbered, old Mirabel is allowed the tag line: "Ads my life Madam, You have got the finest built Watch there, *Tompions* I presume."

The Citizen of the World is less fortunate. In "Letter VIII," he takes home one of "those well disposed daughters of hospitality" with whose kindness he is impressed: "Her civility did not rest here; for at parting, being desirous to know the hour, and perceiving my watch out of order, she kindly took it to be repaired by a relation of her own, . . . and she assures me that it will cost her nothing." We are hardly surprised to learn in "Letter IX," that the little episode cost Goldsmith's Citizen his watch.[21]

Defoe's novels are all written as journals and profit from the immediacy of being told in the first person singular. With the exception of the ending to *Roxana*, they are all success stories giving their readers a vicarious pleasure from the industry of the common man. *Robinson Crusoe* and *Captain Singleton* deal with exotic adventure and are not involved with watches, nor is that part of *Colonel Jack* that takes place in America. Roxana received a gold watch from her "husband"[22]—as Richardson's Pamela receives a "fine repeating-watch" fifteen years later from Mr. B.—but it is for the pickpockets Colonel Jack and Moll Flanders, who want to become gentlefolk, that watches are so important. Moll always felt safer when she was herself dressed up with a gold watch. Before stealing from Lady Betty, she assures us, "I was well dressed, and had my gold watch as well as she."[23]

When he was an apprentice thief, Jack attacked a "Doctor of Physick and a Surgeon" with Will. They came away with two watches as well as other valuables. Later, Will and an accomplice robbed a coach containing a gentleman and a whore. The latter cursed them for taking the gentleman's money and watch, which she considered to be her own perquisites.[24] When they were working their way across country, Jack's accomplice produced a

William Hogarth, *Miss Mary Edwards*, ca. 1740. Miss Edwards' pose suggests the importance of a watch among the adornments of a lady. Swift said that the chains used to shackle Gulliver in Lilliput were "like those that hang to a lady's watch in Europe, and almost as large." (*Copyright The Frick Collection, New York.*)

gold watch that he had taken in a church from a lady's side. Jack was "amaz'd at such a Thing, as that in a Country Town." As they left, they could hear a public reward of ten guineas being offered for the stolen article. There is a similar glimpse of country life in Aubrey's story about a watch being thought the devil, and this may be compared with the request for a plain silver watch in *Memoirs of the Verney Family during the Seventeenth Century*. The plain watch is wanted for timekeeping only, since the lady is certain that no one in the village will have "ather clock or woch."[25]

I have demonstrated elsewhere that although *Moll Flanders* appears to be written in 1683, in that lady's seventieth year, it does in fact deal with a life concurrent with Defoe's (1660–1731),[26] himself born at the beginning of the horological revolution. Moll is motivated throughout by her desire to become a lady. Her bourgeois attention to detail requires her to list her assets (including watches) throughout the novel. At the time when her capital is down to five hundred and forty pounds and she is worried about "that frightful state of life called an old maid," Moll possesses only one gold watch. When she and James discover that they have tricked one another, he is gentlemanly enough to give her what little he has including his gold watch. It is well to remember that Moll did not descend to stealing until she was no longer able to earn her living as a mistress or as a wife. But when she did, she "grew as impudent a thief and as dextrous as ever Moll Cutpurse was." Defoe, through Moll, teaches his readers the useful art of "taking off gold watches from the ladies' sides."[27]

After Charles II introduced the waistcoat in 1675, men wore their watches in a pocket rather than suspended round the neck. This reinforced a tendency for watches (possibly in deference to the Puritan movement) to become plainer after about 1625. After about 1660 the more ornate watches were those intended to be worn by women.[28] There were numerous newspaper advertisements offering rewards for lost and stolen watches, particularly Tompions. The following examples typify an experience of the age: "Taken from a Lady's Side"; "Dropt or taken from a Gentleman's Side"; "Taken from a Gentleman in a coach"; or "Taken away by 6 Highway-men."[29] Until he was hanged at Tyburn in 1725, Jonathan Wild made a living both out of returning such stolen property and as a thieftaker. Defoe wrote the *Life and Actions of Jonathan Wild* in the same year; Wild is also depicted in Gay's *Beggar's Opera* (1728) and Fielding's *Jonathan Wild the Great* (1743).

The very value of the watch made it an insuperable temptation for members of London's low life. Men were generally relieved of their watches by highwaymen or whores; women's watches were fair game for cutpurses of either sex. Moll Flanders' governess, the fence, allowed her no less than twenty pounds for the first watch. The servant girl who eventually caught Moll, before she subsequently turned honest, earned the typical wage of three pounds per year.[30]

Moll obtained her valuable gold watches by a whole spectrum of trickery. A fine watch came from a gentleman to whom she was playing whore in a coach. It was later redeemed by him for thirty guineas. The lady who helped Moll in this transaction kept a "sham gold watch" which she was adept at substituting while a gentleman "was busy with her another way." Even when the victim has fastened "her watch so that it could not be slipped up," Moll knows how to cry "A pickpocket" in order to escape suspicion. One of Moll's gold watches came from her offering to safeguard a lady's goods after a fire, another by her pretending that she knew the mother of the victim, and yet another from the smaller pickings at Bury Fair. As Moll said of her first teacher and herself, before the former was executed, "We had at one time one-and-twenty gold watches in our hands."[31] The teacher's arrest, during "an attempt upon a linen-draper in Cheapside," foreshadows Moll's almost similar fate. Defoe himself had been a hosiery merchant in Cornhill at the time of his marriage in 1683–84—he knew his topic.

The novels offer a mixture of practical advice and moral teaching coupled with a vicarious titillation from material success. There is a rough justice in the fact that Moll has virtually the same amount (seven hundred pounds and two gold watches) at two crucial points in her life. These are the point at which she feels that she ought to stop thieving because she has made enough and the time, much later, when she is obliged to start a new and successful honest life in Virginia at the age of sixty-one. The two gold watches are of particular symbolic and material importance. The one is her badge as a gentlewoman; the other she gives to her son, in America, that her own brother had incestuously sired. Even at London prices, "it was not much less worth than his leather pouch full of Spanish gold" that he gives her in return; but "it was worth twice as much there."[32] London watchmakers were producing a valuable and highly saleable product.

Though a great writer, Defoe is a prosaic one. Watches enter his world because they are an object much to be desired, the most valuable and important domestic artifact in the life of his age. Defoe is little concerned with the poet's traditional theme of the passage of time; still less, in his novels, is he concerned, like the philosophers of his day, with the clock as a potential model for explaining organic life, the universe, or God. But the materialistic regard for watches during the horological revolution, exemplified by Moll, is nevertheless one of the important elements in the influence that watches had on eighteenth-century life and letters.

William Hogarth: The Influence on Art

The artist most directly concerned with the artifacts for measuring time was not a poet but a caricaturist and painter. Hogarth's, life (1697–1764) runs concurrently with the latter part of the horological revolution, so he was suited to portray its influence on society. The works of James Gillray (1757–1815) and Thomas Rowlandson (1757–1827), who donned the mantle of Hogarth, differ in nothing so much as the almost total absence from their works of clocks, watches, and sandglasses.[33] Much the same is true of Joseph Wright of Derby, "the first professional painter directly to express the spirit of the industrial revolution."[34]

There had of course been illustrations of clocks and watches before the time of Hogarth. But these had been exceptional cases rather than themes found throughout the whole corpus of an artist's work. For example, while clocks, watches, and sandglasses are absent from most of Dürer's works, the hourglass is not only portrayed but it is essential to the symbolism of his three most famous copper engravings, produced at the height of his career: *Knight, Death, and the Devil*, *St. Jerome in His Study*, and *Melancholia I*. They are signed and dated 1513, 1514, and 1514 respectively. In all three, the hourglass is prominently displayed above and just to the left of the protagonist. (Dürer's hourglasses have a device above them for marking off the time.) The only other sandglasses I have found in Dürer's work are in *Lansquenet and Death* (1510) and in *St. Jerome in his Cell* (1511). They are clearly forerunners of the great engravings *The Knight* and *St. Jerome* respectively; in each the hourglass is similarly placed.[35] The hourglass after its invention early in the Renaissance was added to the iconography of Time and Death.

Like Dürer, Hogarth was trained as an engraver, but the nature of the times had added a new dimension to that trade. As George Vertue reported of Hogarth, he was "bred up to small engravings of plate work & watch workes."[36] Hogarth seems to have been conscious of the value of time for planning his own work. In his later portraits, he "sometimes painted little more than faces," and proposed "to paint a Portrait in four sittings, allowing only a quarter of an hour to each." Hogarth extended his work study methods to the sale as well as the painting of pictures. Vertue reports on his auction of paintings, which included *A Harlot's Progress*, *A Rake's Progress*, and *The Four Times of the Day*, "by a new manner of sale. . . . to bid Gold only by a Clock, set purposely by the minute hand—5 minutes each lott . . . and by this suble means. [*sic*] he sold about 20 pictures of his own paintings for near 450 pounds in an hour." When Hogarth is defending himself from the suggestion that he is vain, he turns to watchmaking for his *exemplum*. "Vanity," he maintains, "consists chiefly in fancying one doth better than one does"; but if a watchmaker claims that he can make a watch as good as any man, and demonstrates that he really can, "the watchmaker is not branded as infamous."[37]

Because of the age in which he lived, Hogarth, unlike Dürer, is generally concerned with mechanical timekeeping. A sermon glass is prominently displayed in *The Sleeping Congregation* and a vertical sundial in the country scene of *Chairing the Member*, but these reflect areas in which such methods of timekeeping continued to prevail throughout the eighteenth century. In Hogarth's work, they are the exception rather than the rule. We have noted how the originals of *A Harlot's Progress* and *A Rake's Progress* were sold "by a clock." Like Hogarth's other great progress, *Industry and Idleness,* they are both concerned, at one point, with the important "low life" occupation of "taking a watch." In plate 3 we see Hogarth's harlot (who like Moll Flanders wanted to be a "gentlewoman") sitting on the edge of her bed with one breast exposed, and holding up a watch. In Fielding's *Covent Garden Tragedy,* as Paulson points out in *Hogarth, His Life, Art and Times,* "Plate 3 is alluded to when Stormandra reminds Bilkum 'Did I not pick a pocket of a watch,/A pocket pick for thee:'"[38] The influence of Hogarth on Fielding is considerable and freely admitted. For example, in *Tom Jones* (2.3), Mrs. Partridge is said to have "exactly resembled" the harlot's maid in plate 3; in *Tom Jones* (3.6), Fielding says that Thwackum "did in countenance very nearly resemble" the Bridewell taskmaster of plate 4, whose whip is raised to the harlot.

In exactly the same structural position as the *Harlot's Progress* (the third plate out of six), the taking of a watch is once again the central motif for the *Rake's Progress.* In this case, the protagonist is the dupe who sits dallying with the inmates of a bordello. *Industry and Idleness* shows, in twelve plates, the very different progresses of two apprentices who start with equal opportunity. In much the same structural position as the *Harlot* and the *Rake,* the idle apprentice is disclosed in bed with "a common Prostitute"; she has stolen watches in front of her (plate 7). The reverse side of life's coin is illustrated in plate 8, "The Industrious Prentice Grown Rich, & Sheriff of London." This is in sharp contrast with plate 9, "The Idle Prentice Betray'd by His Whore, & Taken in a Night Cellar with His Accomplice." They are caught red-handed; the watches are between the men in the front center of the plate.

But Hogarth goes beyond Defoe; his horological allusions are not limited to the taking of a watch. In addition, he uses clocks for both denotation and connotation. In terms of denotation, *Four Times of the Day* indicate both morning and noon by means of a clock. In much the same way, *The Battle of the Pictures* uses a clock to denote the time of Hogarth's auction, to which reference has already been made. *The Battle of the Pictures*—whose theme is reminiscent of Swift's *Battle of the Books*—was to be used as the ticket of admission "on the last Day of Sale."[39] In *Masquerade Ticket* (second state), the large clock at top center, showing 1:30 in the morning, serves a comparable purpose.

The clock face in the top left corner of *The Times* (hanging outside the home of the government) seems relatively innocuous in the scene of fire and chaos. But *The Times* takes on new connotations when one has "read" the engraving across to the bottom right hand corner. There a destitute child is playing with an almost indentical clock. In the third of the four *Stages of Cruelty*, the gruesome night scene in the graveyard is emphasized by the clock. The woodcut (though not the engraving) has enough lettering underneath for one to decipher the message "memento mori." The full clock shows only in the engraving. The painting and engraving of *Morning* show the same difference between a full and a half clock. In *Southwark Fair*, the clock in the clocktower at the center is pointedly cut in half by the large picture of the Trojan Horse.

Hogarth is, above all, a producer of character portraits; for these, too, he can make use of the watch. In *Analysis of Beauty*, plate 2, the weakness of the apparently cuckolded husband, at the ball, is emphasized by the way that he points to his watch; in the drawing of *Thomas Morell*, the clock above the head of the protagonist serves a similar purpose to that of the hourglass above Jerome's head in Dürer's *St. Jerome in His Study*. The hourglass could be used as a general symbol of Temperance, Time, or Death. But the clock—both through the precise time that it denoted, and the extra possibilities for symbolism that it offered—provided Hogarth with a tool unavailable to his predecessors, and never fully exploited by subsequent artists.

The denotation and connotation of a clock are exemplified in Hogarth's delightful study in seduction, *The Lady's Last Stake*. The clock on the mantlepiece is just about to show sunset; at 4:55, with the moon rising through the window, there is very little time left for the lady's decision. Hogarth describes the subject of the painting as "a virtuous married lady that had lost all at cards to a young officer, wavering at his suit whether she should part with her honour or no [*sic*] to regain the loss which was offered to her." The clock adds to the piquancy of the

William Hogarth, *The Lady's Last Stake*, 1758–59. The lady's titillating dilemma is whether or not she should sacrifice her honor in a last-minute attempt to retrieve the money that she has lost at cards to a young officer. As elsewhere, Hogarth uses the clock effectively for both denotation and connotation. (*By Courtesy of Albright-Knox Art Gallery, Buffalo, New York, Gift of Seymour H. Knox.*)

situation; Cupid with his scythe is mounted above it on a pedestal inscribed "NUNC NUNC." One cannot help but savor the lady's titillating dilemma between the opposing forces of honor and opportunity.[40] Some seventeen years earlier, Hogarth had painted *The Graham Children* with Cupid and his scythe similarly (and perhaps even prophetically) standing above the clock.

An even more ornate clock than the one in *The Lady's Last Stake* stands above the head of the dissipated husband in *Marriage à la Mode*, plate 2. Even without considering the specific symbolism of the clock, one can readily observe how the overdressed man and the overornate clock emphasize each other's excesses. Hogarth's *A Midnight Modern Conversation* was published in two states, and it has been noted that by comparing the two pictures one can see how the artist moved from a portrait group to a picture with moral overtones.[41] The same point is further stressed by the change in the position and nature of the clocks. The relatively small bracket clock on the right of the picture in the first state becomes a towering grandfather clock in the subsequent version. From the left hand rear corner of the room, it unmistakably points out the lateness of the hour to revellers and readers alike.

Not surprisingly, there is also a grandfather clock at the rear of Hogarth's frontispiece for *Tristram Shandy*. This has a symbolic value of its own with which we shall be later concerned. Suffice it to say that Sterne's clock (among other things) already symbolizes some of the negative qualities that came to be associated with the mechanical aspects of clockwork. Hogarth's own awareness of a negative quality in clockwork is demonstrated by his emphasis on the stiff and mechanical attributes of Vaucanson's duck, when he refers to this famous automaton in the *Analysis of Beauty* (1753).[42]

In the same year as the frontispiece for *Tristram Shandy*, Hogarth produced his provocative *The Cockpit* (1759). Paulson perceptively relates this to the influence of Dante and the circular structure of the *Inferno*. Over the very center of this possible picture of hell there falls the highly symbolic shadow of a condemned man holding up a watch. The haunting illustrations that John Martin made for the hell of *Paradise Lost* are a valuable demonstration of the influence of contemporary iron mines, iron works and railway tunnels during the industrial revolution.[43] It is possible that Hogarth used the watch as a comparable metaphor for hell in his *Cockpit*. Certainly, he makes an ominous statement about time in the *Tailpiece*.

Hogarth used a complex symbolism of time for his last and perhaps most haunting and thought-provoking work, *Tailpiece, or the Bathos* (1764). The title of the engraving, *The Bathos, or Manner of Sinking in Sublime Paintings* makes clear the debt to Pope's *Peri Bathous: Or of the Art of Sinking in Poetry* (1727). The subject cannot help but remind one of the end of *Dunciad* 4 (1742–43), when the corruption in the arts has polluted all phases of existence: chaos is come again, "And Universal Darkness buries All."

Hogarth's tailpiece, also produced at the end of his life, relies heavily on symbols of time. Father Time (in the middle of the picture) rests on a broken column. His predecessor had appeared three years earlier in *Time Smoking a Picture*. But now the word *Finis* is written in the smoke that comes from Time's mouth after he has removed his broken pipe. Among the debris lying around are a broken palette, musket, crown, and bottle. The last page of a play shows the words *Exeunt Omnes*; a statute of bankruptcy—sealed with the rider on a white horse (from Revelation)—indicates that Nature is bankrupt; and a flame is just about to consume a picture entitled *The Times*. In the hand of Father Time lies his last will and testament; he bequeathes his world, "all and every Atom thereof to . . . (an erased Lacuna) Chaos whom I appoint my sole Executor." The witnesses, whose seals are affixed, are the three fates, Clotho, Lachesis, and Atropos. Hogarth's message is further stressed by the broken building, the ruined tower, the gibbetted man, the ominous gravestone, the sinking ship, and the falling inn sign that is entitled "The World's End" and shows an orb in flames.

All these are symbols indicating the passage of time. But there are also precise allusions to time reminiscent of the roles that the sun, the moon, the seasons, the bell, the hourglass, and the clock had played in their respective contributions to time measurement. In Hogarth's *Tailpiece*, Phaeton's chariot is falling from the sky, the moon is overcast, the autumnal scythe of time is broken, the great bell is cracked, the sandglass is splintered, and the clock has no hands. In the original drawing, Time had been leaning against a much larger clock face which came between himself and the gravestone. Time's wing partly covered the clock, and the clock partly covered the skull and crossbones at the top of the gravestone. Like Pope twenty years earlier at the end of *Dunciad* 4, Hogarth is making a remarkably pessimistic final statement about the death of Time which conflicts radically with so-called eighteenth-century optimism.

It is difficult to suggest why Hogarth is the only graphic artist of the first rank to have demonstrated so wide an interest in mechanical clocks. One might argue that until the horological revolution clocks and watches were neither as numerous, as accurate, nor as readily marketable as Hogarth's symbolism required. It might also be argued that after the horological revolution mechanical timekeeping lost some of its topical appeal. In addition, as we shall see in due course, certain connotations—like "mechanical art" and "clockwork automatons"—were to put a different emphasis on the clock than is generally evident in Hogarth. But one or two other Hogarths might have been similarly occupied during the period 1690–1760. The debate is comparable to that between two schools of industrial historians. The "heroic" school feels that each inventor has an individual genius which can advance or retard economic history; their opponents maintain that inventors spring out of the matrix of contemporary conditions.[44]

The argument is one for which much may be said on both sides; but there can be no doubt that in art, the similes and metaphors that writers use are necessarily restricted to the artifacts and ideas that they and their audiences share. In addition, analogies become more effective when—like computers, space rockets, or mushroom clouds—they refer to fascinating aspects of current technology. That is why, in part 1, we have first demonstrated how important a revolution occurred in English horology during the seventeenth and eighteenth centuries, and have then indicated the widespread influence of that revolution on both industry and society.

During the horological revolution of 1660–1760 most major Western philosophers, natural philosophers, and theologians (as well as many minor ones) used clock metaphors to explain concepts central to their concerns. But after about 1760 a change is noticeable. Philosophers and scientists tend to prefer biological analogies. Theologians, however (at first so hard to convince), support the argument from design, based on the Watchmaker God and his related clockwork universe, right up to the time of Darwin's *Origin of Species*. In our following two chapters, we will consider first, the use of the clock metaphor by philosophers and, second, the rise and fall of the Watchmaker God.

PART II

The Clock Metaphor in Philosophy and Theology

Philosophers and the Clock Metaphor

Bacon and Descartes

In the seventeenth century, the main thrust of scholarly inquiry came from the moderns, who were particularly influenced by Baconian experimental philosophy and Cartesian mechanistic philosophy. Three centuries later, we may suspect that they were sowing the dragon teeth that our children will have to harvest, but it is hard to detect a premonition of this in their writings. There was a vacillating courage in breaking away from Aristotelianism and the past, combined with a constant effort to retain a place for God in the new philosophy. However, one cannot read even the more pedantic authors of that century without feeling that a revolution was taking place, and "Bliss was it in that dawn to be alive." Out of the scientific and philosophical revolutions came both the horological revolution itself and the metaphor that placed even God in the role of watchmaker.

Bacon's influence raised immeasurably the importance of practical research and the mechanical arts. One of the less obvious attractions of his method was the stress laid on a joint effort in collecting data: "To resolve nature into abstractions is less to our purpose than to dissect her into parts; as did the school of Democritus."[1] Bacon's experimental philosophy had important social implications. He is, in many ways, the originator of the modern society in which everyone can feel that he has talent. Bacon says of his method, that was to provide untold work for many men: "My way of discovering sciences goes far to level men's wits, and leaves but little to individual excellence; because it performs everything by surest rules and demonstrations."[2]

Bacon compares his method to a machine. It is itself a mechanical operation for developing science and technology in his utilitarian spirit. In great measure, democracy has proved to be the godchild of the Baconian method. Science and technology provided the tools for mass production. But mass production required that more and more of the workers become also large-scale consumers. In simplified terms, this is the key to the everwidening political, financial, and social franchise from Bacon's time to our own. Bacon's utilitarian spirit is demonstrated in Solomon's House where experiments are carried out to discover "things of use and practice for man's life."[3]

The Baconian desire to document all practical knowledge was essential to the development of technology. It leads ultimately to the encyclopedias and technological literature that proliferated after the eighteenth century, but more immediately to the experiments associated with the Royal Society. John Webster—a chaplain in the parliamentary army who had earlier been a Cambridge scholar—asks, "Can the *Mathematical* Sciences, the most noble, useful, and of the greatest certitude of all the rest, serve for no more profitable end, than speculatively and abstractively to be considered of?"[4] Science must be put to practical purposes. Elsewhere, after praising "our learned Countrey-man the Lord *Bacon*," Webster combines this call for applied science with the Puritan notion that manual labour has a value in its own right. He is anxious "that youth may not be idlely trained up in notions, speculations, and verbal disputes, but may learn to inure their hands to labour . . . which can never come to pass, unless they have Laboratories as well as Libraries, and work in the fire, better than build castles in the air."[5]

The words of this Puritan divine indicate one reason why Bacon's method appealed to those who persistently attacked idleness as a social and religious evil. It is only more recently that a large segment of youth, for reasons altruistic or otherwise, has attempted to change a long established pattern.[6] While it is true that the Baconian method of testing as much as possible through experiment might ultimately lead to the questioning of God, Bacon was always careful to try to separate science from theology. He stated more than once that while slight knowledge leads to atheism, deeper knowledge leads back to belief in God.[7]

The Baconian movement consciously encouraged a rise in the status of the "mechanical arts." The new attitude towards manual skills (which is still evolving) undoubtedly contributed to God taking on his temporary role of watchmaker. Bacon's preference for the active over the contemplative life points forward to the modern age. He condemned the "pernicious and inveterate habit of dwelling on abstractions," preferring "to begin and raise the sciences from those foundations which have relation to practice, and to let the active part itself be as the seal which prints and determines the contemplative counterpart."[8] But Bacon's position was ambivalent. There is included in the description of Solomon's House a warning that certain experiments should be kept from the public and the state; Bacon's story of "Daedelus; or the Mechanic" in *De Sapienta Veterum* demonstrates a potential for both good and evil.[9]

Bacon's work led naturally into that of the Royal Society. In his *History of the Royal Society*, Sprat underlines the new attitude favoring mechanical arts. He praises artisans and mechanics, while showing a marked lack of attention to what we would call theoretical science.[10] Sprat even goes so far as to hint at an education devoted to the sciences rather than the humanities—at that time, a radical suggestion. Bishop Sprat's attitude is somewhat surprising, since a utilitarian education had been one of the favorite ideas of the discredited Puritans. The Puritan position resulted partly from an attempt to rid the schools of their subjection to Aristotle. Sprat wonders "whether this way of *Teaching* by *Practise* and *Experiments*, would not at least be as beneficial, as the other by *Universal Rules*? . . . In a word, Whether a *Mechanical Education* would not excel the *Methodical*?"[11]

Relationships between scientist and mechanic were of great importance to the horological revolution. Such relationships as those between Huygens and Coster or Hooke and Tompion alert us to the fact that the scientists were developing a special interdependence with the "mechanical arts."

Robert Boyle's preamble to *Some Considerations Touching the Usefulness of Experimental Natural Philosophy* reveals this relationship. Boyle first points out that those presently attempting to set down a record of the mechanical arts have not yet acquired sufficient skill; natural philosophers (or scientists) and virtuosi should be able to convert such descriptions into practical experiment. Boyle feels that he can do better. He is not conceited; almost unbeknown to himself, Boyle has developed what we would now call "professional integrity." He is aware of the novelty of his attempt to set down clearly and precisely the techniques involved in the mechanical arts, "having contented myself to set down such practices faithfully, as I learned them from the best artificers (especially those of *London*) I had opportunity to converse with."[12]

In his apology for science, Boyle stresses how much tradesmen have gained through reciprocal arrangements with scientists. His apology comes to its climax by demonstrating the great advantage accruing to clockmakers through Huygens' recent discovery of pendulum clocks.[13] The language of the moderns seems a little antiquated, and the tone naive because the concepts were then new. But we are witnessing the birth of today's interrelationship between industry and research.

The call by men as different in temperament as Petty, the professor of anatomy at Oxford, and Webster, the Puritan divine, for a close attention to the mechanical arts indicates the improvement in the status of tradesmen. Naturally, this did not occur without some reaction in a class-conscious country like England. In America—where Puritan tendencies have been less held in check by a self-perpetuating conservative aristocracy—Willie Loman would be able to say with conviction that a man who does not know how to handle tools is disgusting. Over a period of time, income and status tend to be related. When plumbers earn more than magistrates or plasterers more than professors, some of the credit (or blame) should go to Bacon and Boyle.[14]

Among the "mechanical artists" who benefitted most from the new science were the clockmakers. Though dilettante experimenters were frequently satirized—in the manner of the treatment afforded to Sir Nicholas Gimcrack in Shadwell's *Virtuoso*—the satire rarely seems to have rubbed off on the true professionals among the "mechanical artists." When the time came for God to be revealed through the metaphor of a watchmaker, the role brought him no loss of status in a mercenary age that was preparing to become mechanically oriented. But it was naturally the thinkers and writers, rather than the mechanical artists themselves, who turned God into a watchmaker.

For the most part, these God-makers belonged to a movement called the moderns. Literary historians have tended to think of the moderns as a predominantly literary movement opposed to classical literature. Since the work of Richard Foster Jones, however, we know that the literary quarrel was little more than the reflection of an earlier and much more crucial battle between the champions of ancient and modern science.[15] Swift, too, points to the central protagonists. In his *Battle of the Books*, dealing essentially with the literary "Querelle," he lets Aristotle shoot at Bacon and Descartes.

Although they are at one in attempting to throw off the shackles represented by Aristotle, Bacon and Descartes differ in championing an experimental and a system-centered approach to natural philosophy, respectively. The analogue of clockwork is essential to the Cartesian mechanistic philosophy because there were no equally appropriate mechanical devices at the time. Though clock analogies had existed at least since Aquinas and even Bacon had used them for animals and the universe, Descartes was the first to incorporate such analogies methodically into a philosophical system. As a result, the Cartesian philosophy is very much mechanically oriented, but Descartes rarely forgot that his ultimate fate lay with God. Though he explains the body in mechanical terms, Descartes—who may very well have read the *Summa Theologica*—differentiates, like Aquinas, between man and beast.

The involvement of Descartes with animal mechanism may well be related to his passion for automata. Long before publishing the *Discourse*, he toyed with the notion of constructing a human automaton. Poisson, one of his correspondents, states that in 1619 he planned to construct a flying pigeon, a dancing man, and a spaniel chasing a pheasant. There is a legend that he did in fact build a beautiful blond automaton named Francine, but that she was discovered in her packing case on board ship and dumped over the side by a captain afraid of witchcraft. There may be no more truth in these rumours than in similar stories about Albertus Magnus and others, but it does at least suggest an early fascination with automata.

What we do know is that this was a practical period in Descartes' life. After graduating in law in 1616, he had become weary of study and wished to see the world. This led him to join the army of Prince Maurice of Nassau, for whom his mathematical ability was useful in military engineering. At Breda in 1618–19, Descartes met Isaac Beeckman, the mathematician, and together they discussed the question of making physical problems amenable to mathematics. In Germany on St. Martin's Eve, 10 November 1619, it was revealed to Descartes in a dream how all sciences are interconnected "by a chain." The new prophet spent the following nine years applying his revelation to algebra.

But the fascination with automata never left Descartes. Their history is interrelated with that of the clock, and it is both natural and convenient that these should provide the twin analogues for Cartesian mechanistic pholosophy.

The Symbol of Order Inherent in the Watch Analogue

There has been much dispute about the relative influence of Bacon and Descartes in England. We shall discover, however, that in the period leading up to and during the horological revolution all of the major philosophers and many of the minor ones use clock analogues irrespective of the schools with which they are associated. This is surely because their analogies are at least as dependent upon a suitable contemporary artifact as upon ideas that they may borrow from one another.

What is more worthy of our attention than a debate over influence is the question of why clock analogues should have taken precedence over all others during the particular period with which we are concerned. Part of the answer is that it was the period during which clocks and clockmakers achieved their greatest prestige. But this is an oversimplification; other inventions, like the telescope and the marine compass, were making a comparable impact on the public imagination. We must look further into the intellectual and social changes that were taking place.

It is a commonplace that the questioning of established patterns of thought and religion during the Renaissance included a questioning of Aristotle, whose teachings had been incorporated into those of the established church. This was accompanied, in part, by a turning towards the atomic theories ascribed to Democritus, Epicurus, and Lucretius. The questioning of a structured feudal world and an even more highly structured Ptolemaic universe brought some very real fears of chaos upon earth.

Religious wars on behalf of and against the Catholic church devastated Europe during the sixteenth and seventeenth centuries. France, the greatest country on the continent, was ravaged by the Hundred Years' War, and was only beginning to recover during the lifetime of Descartes. The founding of the Académie Française together with the classical restraint preached by Malherbe and Boileau are merely pointers in the literary field to a much more general desire for order which followed the war. A widespread acceptance of the need for order does much to explain not merely the *unités* and *règles* in literature, the classical lines of architecture, and the symmetrical patterns of gardens, but also the autocracy of the Roi Soleil and ultimately even the reaction which led to the revocation of the Edict of Nantes. By influencing Huygens' move to Holland, by encouraging the exodus of Huguenots, and by helping relations between the English and the Dutch, this French decision contributed to the rise of English horology.

A comparable pattern of religious warfare after a breaking down of the old structure also took place in Germany; their Thirty Years' War ran concurrently with Descartes' adult life. In England, the major clash came a little later, during the twenty years before 1660. As in France, this was followed by a classical period, and for much the same reason. But Augustan values are the exception rather than the rule in English literature. The movement proved neither as rigid nor as homogeneous as classicism in France.

During the seventeenth century, then, men were consciously seeking a formula for order. They desperately needed a new way to understand the universe and man, something to replace the old systems yet to remain, like them, an antidote for chaos. This made the symbol of order inherent in the watch analogue particularly appealing.

In Fontenelle's popular interpretation of Descartes what attracts the countess about the universe's being like a watch is the fact that it makes "more plain and easie" the "whole *order* of Nature" (italics added).[16] When Boyle is writing of his corpuscles, he too is attracted by the order that he finds in them. He compares their movement to that of the great clock at Strasbourg in which the "parts of it, move several ways... as *regularly* and *uniformly* as if it knew and were concerned to do its duty." Later, he says significantly that what attracts him about a clock is "how *orderly* every wheel and other part performs its own motions" (italics added).[17] Just as Boyle's clock provides an orderly model for corpuscles, Hobbes bases his ordered model of society in the *Leviathan* on an analogy with a watch. But when Cudworth attacks Hobbes' materialism, he too uses the analogy with a watch. For Cudworth, "*Order* and proportion" (italics added) in a watch are "in the Intellect it self" rather than in the watch or in the eye that perceives it.[18]

So compelling was the desire for order that even those who accepted the atomic theory rejected its potentially atheistic concept of atoms coming together through chance. McFarland demonstrates well that although the ideas of Democritus, Epicurus, and Lucretius were in varying degrees acceptable in their modified form, it was precisely with the element of chance that thinkers like Bacon, Newton, Voltaire, Hume, and Kant found fault.[19]

For the corpuscular model modified to seventeenth-century requirements, the watch provided the new and perhaps essential analogue. Lord Herbert of Cherbury complains that Epicurians "attribute all things to Chance." He uses a watch to demonstrate the quality of order in nature: if anyone "observe a Watch, shewing the hours exactly" for a day, he will conclude that it was made by an artisan; how much more so will a person "who does but contemplate the vast *Machine* of this World, performing its motions so regularly," conclude that it was made by "an all Wise and Powerful Author."[20] The quality of order in the watch, as in the world, was essential. In the preface to Nieuwentyt's *Religious Philosopher*, there is later developed the same analogue. The author asks rhetorically whether a man "can perswade himself that this Watch . . . could have acquired its Being and Form by mere Chance only . . . and without any certain Rule or Direction?"[21]

Writers as different as Mandeville and Ray are at one in condemning the "fortuitous" element in classical atomism. Mandeville says, "The Doctrine of *Epicurus*, that every thing is deriv'd from the Concourse and fortuitous Jumble of Atoms, is monstrous and extravagant beyond all other Follies."[22] John Ray feels that Cudworth has sufficiently refuted the "whole *Atomical Hypothesis*, either *Epicurean* or *Democritick*," but "cannot omit the *Ciceronian* Confutation" in *De Natura Deorum*: "Such a turbulent Concourse of Atoms could never . . . compose so well order'd and beautiful a structure as the *World*."[23]

Philosophers, even more than poets, seem to stress those elements in the clock analogue that underline order. They are less involved with fortune being like the swing of a pendulum, passions being wound up, or life riding on a dial's point. In general, it is with the total watch and with the impressive order of its wheelwork that they are concerned.

As early as 1377, Oresme likened the heavens "to a man making a clock and letting it run and continue its own motion by itself." A few lines further on, he adds that God "ordained and deputed angels who should move the heavens and who will move them as long as it shall please Him."[24] Oresme is foreshadowing (at a distance of three and a half centuries) the celebrated controversy between Leibniz and Samuel Clarke the Newtonian. The watch universe of Leibniz demonstrated a "beautiful preestablished *order*" (italics added). Leibniz felt that his system was better regulated than that of the opposition because "according to their doctrine, God Almighty wants to wind up his watch from time to time."[25]

Bacon himself, though less concerned with hypotheses, was well aware of the analogy between clocks and both the universe and animals. In the *Novum Organum*, he maintains that "the making of clocks... is certainly a subtle and exact work; their wheels seem to imitate the celestial orbs, and their alternating with *orderly* motion, the pulse of animals..." (italics added). It is also natural that Descartes, who was so concerned with the clockwork nature of animals, would use the same analogy for the universe. In *Principles* 4.228, he says that he has "described the earth, and all the world that is visible, as if it were simply a machine...."

Baconians employed the concept of the universe being like a watch just as readily as did the followers of Descartes. Wilkins suggests an analogy between watches and the universe in book 2 of *A Discourse concerning a New Planet*, first published anonymously in 1640: "Wee allow every Watch-maker so much wisdome as not to put any motion in his Instrument, which is superfluous, or may bee supplied an easier way: and shall wee not think that Nature ha's as much providence as every ordinary Mechanicke? Or can wee imagine that She should appoint those numerous and vast Bodies, the Stars to compasse us...."[26] Power's Language is permeated with such mechanical metaphors as "the Crystalline wheelwork of the Heavens,"[27] and Boyle compared the universe to the great clock at Strasbourg.[28] Frequently, of course, the analogy no longer has to be explicitly made. Bolingbroke, like others, can refer to the universe as "this vast machine."[29]

Apart from demonstrably replacing an old with a new sense of order in the universe, the clock analogy helped to reduce some of its mystery. This is also true of the use of the clock analogy for explaining the "mechanism" of animals. Indeed, dissection and even vivisection of animals is closely connected with the mechanistic philosophy from Descartes onwards.

Descartes and the Mechanistic Philosophy

Descartes was plagued by the need to differentiate between men and animals. Man has a soul; but more than this is essential for the demonstration of his superiority as given in the *Discourse on Method.* First Descartes maintains that a beast cannot communicate its thoughts through language and second, that a beast lacks reason. He feels that even the most unintelligent of men can reply logically to questions, and in addition "it is morally impossible to have enough different organs in a machine to make it act in all the occurrences of life in the same way as our reason makes us act." Descartes is equating animals with machines (he says elsewhere, "I recognize no difference between these machines and natural bodies"), but he insists that a man is something different. Descartes then reinforces his case through an analogy which appears to relate clockwork to orderly but unreasoned motion:

> . . . although there are many animals which show more skill than we do in certain of their actions, yet the same animals show none at all in many others, so that what they do better than we does not prove that they have a mind. . . . as one sees that a clock, which is made up of only wheels and springs, can count the hours and measure time more exactly than we can with all our art.[30]

There can be no doubt about Descartes' differentiation between man and animals. He states clearly that one should not think, as did some of the ancients, that animals do speak, although we do not understand their language. The final paragraph of "Discourse 5" is particularly revealing in respect to the human soul: "There is nothing which leads feeble minds more readily astray from the straight path of virtue than to imagine that the soul of animals is of the same nature as our own, and that, consequently, we have nothing to fear or to hope for after this life, any more than have flies or ants. . . ."[31] Viewed from the comparative objectivity of a distance of three and a half centuries, one can sympathize with Descartes' agonizing dilemma between a mechanical interpretation of organic life, and the very real need to provide a place for the human soul.

In the year following the publication of the *Discourse*, Descartes supported his thesis on the difference between animals and men by using an analogy with automata. The argument is somewhat involved, but the introduction demonstrates Descartes' problem:

> Most of the actions of animals resemble ours, and throughout our lives this has given us many occasions to judge that they act by an interior principle like the one within ourselves, that is to say, by means of a soul which has feelings and passions like ours. . . .

Descartes then argues that if a man, who had previously seen automata but not animals, ever saw real animals, or "automata made by God or nature to imitate our actions," he would still recognize them as being automata rather than men. Descartes' reason for this is that automata and animals can neither speak logically nor act "as our reason makes us act." The passage concludes: "We base our judgment solely on the resemblance between some exterior actions of animals and our own; but this is not at all a sufficient basis to prove that there is any resemblance between the corresponding interior actions."[32] The last argument was made through a clock analogy that will receive separate attention.

Some eleven years later, Descartes was involved in correspondence with Henry More, the English philosopher and poet. At that time, he expressed the same views regarding the analogy with automata and the difference between man and animals: "It seems reasonable, since art copies nature, and men can make various automata which move without thought, that nature should produce its own automata, much more splendid than artificial ones. These natural automata are the animals."[33] Just as animals are like automata, Descartes feels that they are, of course, also like clocks. None have mind or soul and therefore a similar analogy holds good:

> I know that animals do many things better than we do, but this does not surprise me. It can even be used to prove they act naturally and mechanically, like a clock which tells the time better than our judgement does. Doubtless when the swallows come in spring, they operate like clocks. The actions of honeybees are of

> the same nature, and the discipline of cranes in flight, and of apes in fighting. . . . To which I have nothing to reply except that if they thought as we do, they would have an immortal soul like us.[34]

In the same letter to Henry More, Descartes defends his position regarding the absence of souls in animals. He argues that if we granted a soul to some animals we would have to grant it to all. But that would include oysters and sponges which are too imperfect for this to be credible. One can suspect that the defensive posture of the mechanistic philosophy resulted less from animals being externally like men than from men being externally like animals.

The place of the soul in Cartesian philosophy would plague thinkers right through the eighteenth century, but the soul's potential anomaly in the mechanistic view of life was evident from the beginning. Even in Holland, the land of toleration, Descartes met with enmity. The bitterest came from Gisbert Voet, president of the University of Utrecht. By insinuating atheism, Voet was able to secure Descartes' condemnation from the local magistrates. Henry More (to whose correspondence with Descartes we have already referred) was at first an admirer of the French philosopher; he came later to think that Cartesianism would lead to a form of mechanical naturalism, and from that to atheism.

Inevitably, one line of philosophical development did lead from the mechanical animal of Descartes to La Mettrie's *L'Homme-Machine* (1747–48).[35] In this period of growing sensibility, others preferred to give souls to the animals rather than take them from men. In either case, the clock analogy played its part.

Mersenne—whom Descartes credited with having more in him than "all the universities together"—took a line similar to that of his friend and former schoolfellow. He felt that beasts are obliged to follow sense impressions: "comme il est necessaire que les roües d'une horloge suivent le poids ou le ressort qui les tire. Mais l'homme . . . remarque et sépare ce qui est de corruptible et d'incorruptible, de muable et d'immuable, de finy et d'infiny dans chaque chose."[36]

The clock analogue's great value was that it "explained"

animal mechanism. Cartesian dualism allowed no soul to animals and kept the beast and the soul separated in man. It was a dualism in which "all reason is spiritual and immortal," and "all matter incapable of thought." In the Cartesian system, "body, and the animal body in particular . . . cannot partake of soul." For Descartes, there were no half measures on this point. He could accept "neither intermediate substance, nor substantial form, nor sensitive soul, all of which were claimed by neo-Aristotelians; nor the portion of the *Anima Mundi* allotted to beasts by the neo-Platonists; nor even the inferior rational soul granted them by seventeenth-century Epicureans and Pyrrhonians."[37]

In his *Treatise of Man*, Descartes uses the clock analogy to demonstrate that he had developed a physiology in which the operations of the flesh were to be considered separately from those of reason, mind, and soul: "ces fonctions suivent toutes naturellement, en cette Machine, de la seule disposition de ses organes, ne plus ne moins que font les mouvemens d'une horloge, ou autre automate, de celle de ses contrepoids & de ses roües, en sorte qu'il ne faut point à leur occasion concevoir en elle aucune autre Ame vegetative, ny sensitive, ny aucun autre principe de mouvement & de vie. . . ."[38]

Later Cartesians

Like Descartes, the German and Dutch Cartesians used the clock simile to explain their mechanistic view of animal life. The concept of "occasionalism" (a "second cause" system which postulates that things are only "occasional causes," God is the sole efficient cause) was systematized by the French Cartesian, Father Malebranche. Clauberg, the German philosopher and theologian, used occasionalism partly to overcome the problem of Cartesian dualism. Clauberg followed Descartes (as well as Donne and Herbert before them)[39] in explaining the difference between a living and a dead body by comparing them respectively to a watch in running order and a watch that has stopped.

Henri de Roy (Henricus Regius)—another Cartesian, and a professor of medicine at the University of Utrecht—was almost condemned in that town as a heretic. In his *Fundamenta physices*, de Roy defends the essential difference between man and beast in terms of the soul. He feels that animals and clocks can be trained to achieve the particular ends that human beings require.[40]

In England, Henry Power—another of the many "doctors of physic" smitten with science—explained how, with the aid of the microscope, the Cartesian mechanical dream was about to emerge as a visible reality. Power is confident that we will be able to "see what the illustrious wits of the Atomical and Corpuscularian Philosophers durst but imagine, even the very Atoms... nay the curious Mechanism and organical Contrivance of those Minute Animals...." Power cannot resist directing a little good-natured raillery at the "vanquished" Aristotle: "Were *Aristotle* now alive, he might write a new History of Animals; for... the Naturalists hitherto... have regardlessly pass'd by the Insectible *Automata*, (those Living-exiguities) with only a bare mention of their names, whereas in these pretty Engines... are lodged all the perfections of the largest Animals.... in these narrow Engines there is more curious Mathematicks, and the Architecture of these little Fabricks more neatly set forth the wisdom of their Maker."[41]

If the corpuscles could not be seen, they might at least be heard. Hooke suggested that there could "be a possibility of discovering the internal motions and actions of bodies by the sound they make" (an interesting variation of the music of the spheres). He feels that if we cannot see the corpuscles with the aid of a microscope, we may be able to hear them, as we hear the internal movements of a watch: "Who knows," he suggests, "but that as in a Watch we may hear the beating of the Balance, and the running of the Wheels, and the striking of the Hammers, and the grating of the Teeth, and the Multitude of other Noises; who knows, I say, but that it may be possible to discover the Motions of the Internal Parts of Bodies, whether Animal, Vegetable, or Mineral, by the sound they make...."[42]

In *A Free Inquiry into the Received Notion of Nature*, Boyle seems to offer a revealing defense of what amounts virtually to a Cartesian approach to animal mechanism: "If it should be disliked that I make the phenomena of the merely corporeal part of the world, under which I comprise the bodies of animals, though not the rational souls of men, to be generally referred to laws mechanical; I hope you will remember... that almost all the modern philosophers, and among them diverse eminent divines,... endeavour to account for what happens in the incomparably greatest part of the universe, by physico-mechanical principles and laws."[43]

Though the Cartesians experienced difficulty in transferring animal mechanism to men, they could readily relate it to lower orders like vegetables. Arnold Geulincx, the Dutch "occasionalist" philosopher, used the clock analogy in his *Ethica* to describe a world in which not only animals but also plants operate like "clockwork."[44] One senses the element of mechanical "predestination" in the clock analogue when Descartes uses it in connection with a tree: "It is certain that there are no rules in mechanics which do not hold good in physics... for it is not less natural for a clock, made of the requisite number of wheels, to indicate the hours, than for a tree which has sprung from this or that seed, to produce a particular fruit."[45]

Sir Kenelm Digby, an English follower of Descartes, attempted to explain the difference between plants and animals through an analogy with two water-driven machines, one simple and one complex, that he saw in Spain during his youth. In relating the complex machine to an animal, he says: "Now because these parts (the movers, and the moved) are partes of one whole; we call the entire thing Automatum or *se movens*; or a living creature,"[46] It is not merely the tortuous language, but the lack of a readily recognizable analogue that spoils Digby's argument. This indicates all too well why—though the magnet and the telescope may have been just as important to the science of the seventeenth century—only the watch could provide its essential analogue for the mechanism and the order in animals, society, and the universe.

The case with man himself was difficult. His soul's potential anomaly in the mechanistic view of life was evident from the beginning. Descartes' theory of the soul operating through the pineal gland was never really satisfactory. He begins article 34 of *Traité des passions*: "Concevons donc icy que l'ame a son siege principal dans la petite glande qui est au milieu du cerveau...." A passage in *De la Formation du foetus* ("une Ame dans une horloge, qui fait qu'elle monstre les heures")[47] indicates the type of image with which he is trying to come to terms. In *L'Homme*, Descartes seems to think in mechanical terms not only of the body but also of the mental processes:

> Je desire que vous consideriez après cela, que toutes les fonctions que j'ay attribuées à cette Machine, comme la digestion de viandes, le battement du coeur & des arteres... l'impression de leurs idées dans l'organe du sens commun & de l'imagination; la retention ou l'empreinte de ces idées dans la Memoire.... Je desire, dis-je, que vous consideriez que ces fonctions suivent toutes naturellement en cette Machine, de la seule disposition de ses organes; ne plus ne moins que font le mouvemens d'une horloge, ou autre automate, de celle de ses contrepoids & de ses roües....[48]

Descartes attempted to retain the soul in his system as something different from the mechanical body and as something unique to man. He hoped, no doubt, that the dualism would satisfy both the church and his own religious qualms. The "occasionalism" or "second cause" system of Father Malebranche produced such equally perplexing paradoxes as the question of how an immutable God could be continually intervening to produce the ever changing face of nature. Like Descartes, Malebranche did not favor the Augustinian view that animals have a soul. He supports his argument with an interesting use of the watch analogy: the order of a watch's wheelwork and the regulation of its movement demonstrates intelligence, but intelligence is not matter, "elle est distingée des bêtes, comme celle qui arrange les roües d'une montre est distinguée de la montre."[49]

Hobbes, on the other hand, took the materialistic philosophy to its logical conclusion. For him, everything including the soul was material: "Every part of the Universe, is Body; and that which is not Body, is no part of the Universe.... Nor does it follow from hence, that Spirits are *nothing*: for they have dimensions, and are therefore really *bodies....*"[50] Hobbes probably attracted more vehement accusations of atheism for this concept than any man of his age.

Coping with the Cartesian "Clockwork" Philosophy

Cudworth, the Cambridge Platonist, produced "the first full-length attack on Hobbes."[51] He was infuriated by "that Prodigious Paradox of Atheists, that *Cogitation* itself is nothing but *Local Motion* or Mechanism...." This heresy, like much else, is attributed to Hobbes: "A Modern Atheistic Pretender to Wit, hath publickly owned this same Conclusion, that *Mind is Nothing else but Local Motion in the Organick parts of Man's Body*." Descartes, he continues, may have suggested that animals are mechanical automata, but he at least felt that they could not think. To suggest that "Cogitation" itself is a material process makes the problem much more serious: "that *Cogitation* it self, should be *Local Motion*, and *Men* nothing but *Machines*; this is such a Paradox, as none but either a Stupid and Besotted, or else an *Enthusiastick*, Bigotical, or Fanatick Atheist, could possibly give entertainment to. Nor are such men as these, fit to be Disputed with, any more than a *Machine* is."[52] Cudworth then accuses the atheists of denying a first cause, and returns to the ultimate heresy that men themselves would be considered machines.

Henry More's situation emphasizes the ambivalent position in which some philosophers found themselves. This Cambridge Platonist was partly responsible for bringing the Cartesian philosophy to England. His correspondence with Descartes (1648–49) still offers one of the best introductions to the Cartesian concept of the animal as a machine. Nevertheless, More repudiated the concept strongly and later turned against Cartesianism altogether in favor of experimental philosophy. The change is marked by his *Enchiridion Metaphysicum* (1671). Cudworth, too, eventually found experimental philosophy closer to his taste. (The Cambridge Platonists had Puritan tendencies that may help to explain this.) He wrote to Boyle: "But your pieces of natural history are unconfutable, and will afford the best grounds to build hypotheses upon. You have much outdone *Sir Francis Bacon*, in your natural experiments. . . ."[53]

Ironically, Bacon could use a mechanical image for the mind with impunity. He contended that "nothing is more politic than to make the wheels of the mind concentric and voluble with the wheels of fortune." But Bacon, like the poets, was saved by not basing hypotheses on such metaphors. Bishop Sprat goes further, in *The History of the Royal Society*, though he carefully qualifies his statement: "For, though Man's *Soul*, and *Body* are not onely one *natural Engine* (as some have thought) of whose motions of all sorts, there may be as certain an accompt given, as those of a Watch or Clock: yet by long studying of the *Spirits*, of the *Bloud*, of the *Nourishment*, of the parts . . . there, without question, be very neer ghesses made, even at the more *exalted*, and *immediate* Actions of the *Soul*; and that too, without destroying its *Spiritual* and *Immortal* Being."[54]

It appears that the fundamental problem of finding a place for the soul had either to be put to one side, rationalized, or squarely faced by every philosopher during the horological revolution. Locke seems to put the problem gently to one side by seeing "what objects our understandings were, *or were not*, fitted to deal with" (italics added);[55] Leibniz—relying on a discovery of Huygens regarding the synchronizing quality between adjacent pendulums—ingeniously suggests that the body and the soul are like two clocks whose harmony is preestablished;[56] and Hobbes maintains that body and soul are both material.

The only book that Spinoza ever published with his own name on the title page was a geometrical version of Descartes' *Principia*. But Spinoza went beyond the dualism of Descartes. In *Ethics*, he says that "the ancients, so far as I know, never found the conception put forward here that the soul acts according to fixed laws, and is as it were an *immaterial automaton*" (italics added).[57] Spinoza—the poor lens grinder who eventually died from pulmonary disease contracted through inhaling glass dust—suffered problems enough on account of his philosophy and religion. But he was to some extent safeguarded by the fact that, as La Mettrie put it, the system was "plein de ténèbres."

After he became involved in his theory of "L'Homme-Machine," La Mettrie changed from denigration to praise his attitude towards Spinoza's mechanism. Two passages deleted from the 1745 edition of La Mettrie's *Histoire Naturelle de l'âme* indicate the way in which Spinoza was liable to be interpreted by contemporary thinkers. In talking of birds, La Mettrie says, from a point of view that was soon to change, that "ce ne sont donc point, encore une fois, des automates, comme le veut Descartes, semblables à une pendule ou au fluteur de Vaucanson. Et à plus forte raison Spinoza a-t-il tort de prétendre que l'homme ressemble à une montre plus ou moins parfaite (qui marque les heures, les minutes, le jours du mois, de la Lune, ou seulement quelques-unes de ces choses, selon son mécanisme ainsi qu'elle les marque plus ou moins régulièrement selon la bonté & la justesse de ses ressorts). . . ."

In another deleted passage dealing similarly with Spinoza, La Mettrie had said: "Ne connoissant ni Dieu, ni Ame, Cartésien outré, il fait de l'homme même un veritable automate, une machine assujettie à la plus constante nécessité, entraînée par un impétueux fatalisme, comme un vaisseau par le courant des eaux." By way of contrast, La Mettrie added to the text of 1751: "Suivent Spinosa encore, l'homme est un véritable Automate, une Machine assujettie à la plus constante nécessité."[58]

But if Spinoza was to some extent protected by being "plein de ténèbres," the two philosopher-physicians La Mettrie and Hartley were not. In the middle of the eighteenth century, they justified fully the fears concerning mechanistic thought that had been expressed from the beginning. Both came as close to atheism as at that time was decently possible.

Hartley developed a mechanical theory of learning through association based on pleasure and pain. Like others before him, he was concerned with the "difficulty of supposing that the soul, an immaterial substance, exerts and receives a physical influence upon and from the Body." He feels that the mind and body "must be related." The former Hartley considers to be controlled by the "Doctrine of Vibrations," for which he gives credit to Newton, and the second by the doctrine of "Association," for which he gives credit to Locke.[59]

Hartley has moved beyond the Cartesian attitude towards

animals. Much of what he has to say, however, reminds us both of that source and of associated ideas that have influenced the "behavioral sciences." He maintains, for example, that: "The *motions* of the Body are of two kinds, automatic and voluntary. The *automatic* Motions are those which arise from the Mechanism of the Body in an evident manner. They are called *automatic*, from their Resemblance to the motions of *Automata*, or Machines, whose Principle of Motion is within themselves." Of animals, Hartley says: "Though I suppose, with Descartes, that all their Motions are conducted by mere Mechanism; yet I do not suppose them to be destitute of Perception, but that they have this in a manner analogous to that which takes place in us; and that it is subject to the same mechanical laws as the Motions." Man, too, is to be understood in mechanical terms: "By the Mechanism of human actions, I mean, that each Action results from the previous Circumstances of Body and Mind, in the same manner, and with the same Certainty, as other Effects do from their mechanical Causes. . . ." What cannot be deduced from observation of the workings of the mind may be believed from analogy, as occurs "when a Person first changes his Opinion from Freewill to Mechanism, or more properly first sees Part of the Mechanism of the Mind, and believes the rest from Analogy. . . ."[60]

Hartley's mechanical philosophy addresses itself to those who would today be called professional philosophers, but La Mettrie's *L'Homme-Machine* (1747–48) is a more popular work. It pursued the clock analogy to its logical conclusion, and became a *cause célèbre*.

As a physician, La Mettrie derives many of his ideas about the mechanical nature of behavior from the observation of patients, but he thinks also in terms of his horological metaphor. In the first page of *L'Homme-Machine*, he tells us that "demander si la Matière peut penser, sans la considérer autrement qu'en elle-même, c'est demander si la Matière peut marquer les heures." For him, medicine consists of regulating the pendulum: "Cette oscillation naturelle, ou propre à notre Machine, & dont est douée chaque fibre, &, pour ainsi dire, chaque Elément fibreux, semblable à celle d'une Pendule, ne peut toujours s'exercer. Il faut la renouveller, à mesure qu'elle se perd; lui donner des forces quand elle languit; l'affoiblir, lorsqu'elle est opprimée par un excès de force & de vigueur. C'est en cela seul que la vraie Médecine consiste." Though the body is (as one would expect) a watch, the watchmaker has become a sort of gastric juice, "Le corps n'est qu'une horloge, dont le nouveau chyle est l'horloger." Understandably, the work concludes along the lines that Nature is like a master watchmaker and man is no more than a machine: "La Nature n'est point une Ouvrière bornée. Elle produit des millions d'Hommes avec plus de facilité & de plaisir, qu'un Horloger n'a de peine à faire la montre la plus composée."[61]

What impresses La Mettrie, above all, are the technical possibilities of his age, and ability to "explain" man in terms of such analogues: "Je me trompe point; le corps humain est une horloge, mais immense, et construite avec tante d'artifice & d'habilitée, que si la roüe qui sert à marquer les secondes, vient à s'arrêter; celle des minutes tourne & va toujours son train; comme la roüe des Quarts continüe de se mouvoir: et ainsi des autres, quand les premières, roüillées, ou dérangées par quelque cause que ce soit, ont interrompu leur marche."[62] He refers to specific inventions that were making some men feel their technology was approaching that of God. In the repeater watch (that particular favorite of the eighteenth century), man's technology seemed to come close to imitating his own rational intelligence. The clockwork automata of Vaucanson, which caused such excitement after their exhibition in 1738, made man feel that he was narrowing the gap between his machines and God's organisms. If, indeed, he could manufacture organisms equal to those of God, the clockwork analogy might die, but man would stand in the place of his Maker.

Materialism is defended by La Mettrie in terms of the very inventions through which his optimism is derived. He says that man "est au Singe, aux Animaux les plus spirituels, ce que la Pendule Planétaire de Huygens, est à une Montre de Julien le Roi. S'il a fallu plus d'instrumens, plus de Rouages, plus de ressorts pour marquer les mouvemens des Planètes, que pour marquer les Heures, ou les répéter; s'il a fallu plus d'art à Vaucanson pour faire son *Fluteur*, que pour son *Canard*, il eût dû en emploier encore davantage pour faire un *Parleur*; Machine qui ne peut plus être regardée comme impossible. . . ."[63]

Today we know that the minds and bodies of even the simplest living organisms are far more complex than Vaucanson's duck or a repeater watch. But, in suggesting that a machine capable of speech should no longer be regarded as impossible, La Mettrie indicated the type of optimism in man's technology that the horological revolution could inspire. A century earlier no one questioned Descartes' insistence that neither automata nor animals might be expected to speak logically, nor act "as our reason makes us act."

Clusters of Clock Analogies

In arguing that animals did not have a human potential, Descartes insisted that any similarity in external actions should not be used to prove a "resemblance between the corresponding interior actions." This concept was supported by a clock analogy. It appears to be the first of a particular cluster of clock analogies, and there are other such distinct clusters or patterns that can be traced through the history of these metaphors. As an example, we have already noted how Donne, Herbert, Descartes, and Clauberg all attempted to explain the difference between a body that is alive and a body that is dead through an analogy with a watch that is going and a watch that has stopped.

With reference to the important cluster of analogies demonstrating that external appearance does not necessarily explain internal mechanism, Descartes has the following example in *Principles*:

> ... bien que *j'aye peut-estre imaginé* des causes qui pourroient produire des *effets semblables à ceux que nous voyons*, nous ne devons pas pour cela conclure que ceux que nous voyons sont produits par elles. Pource que, comme un horologier *industrieux* peut faire deux montres qui marquent les heures en mesme façon, & entre lesquelles il n'y ait aucune difference en ce qui paroist à l'exterieur, qui n'ayent toutefois ... rien de semblable en la composition de leurs roües: ainsi il est certain que *Dieu* a une infinité de divers

> moyens, par chacun desquels *il peut avoir fait que* toutes les choses de ce monde *paroissent telles que maintenant elles paroissent*....[64]

Bougeant also recognizes that one cannot know the internal machinery from the external appearance. Though he would be loath to believe that his friends are "nothing but Machines," he knows that "God has the power of making such creatures as should have the Appearance only and Motion of Men, though they were at bottom nothing but Machines,"[65]

Glanvill, like More by whom he was influenced, tended to move from Descartes in the direction of the new experimental philosophy. He hesitates to accept the Cartesian system as more than hypothesis:

> And though the Grand Secretary of Nature, the miraculous *Des-Cartes* have here infinitely out-done all the Philosophers went before him...he intends his principles but for *Hypotheses*,...For to say, the *principles* of Nature must needs be such as our *Philosophy* makes them, is to set bounds to Omnipotence, and to confine *infinite power* and *wisdom* to our *shallow models*....we can have no true knowledge...except we comprehend all....Thus we cannot *know* the cause of any one *motion* in a *watch*, unless we were acquainted with all its motive dependences, and had a distinctive comprehension of the whole *Mechanical* frame.[66]

Some twenty pages earlier in *Scepsis Scientifica*, Glanvill uses a clock analogy which, like that of Descartes, relates exterior appearance to internal mechanism. But Glanvill uses the analogy to emphasize the value of experimental philosophy *vis-à-vis* the Aristotelians: "Nature is set going by the most *subtil* and *hidden* instruments, which it may have nothing *obvious* which resembles them. And therefore what shews only the outside, and sensible structure of Nature; is not likely to help us.... 'Twere next to impossible for one, who never saw the inward wheels and motions, to make a watch upon the bare view of the *Circle of the hours*, and *Index*:...For *Nature* works by an *Invisible Hand* in all things: And till *Peripateticism* can shew us further...never make us Benefactors to the World, nor considerable Discoverers."[67]

Power's optimism about the new science is greater: "Who can tel how far Mechanical Industry may prevail; for the process of Art is indefinite, and who can set a *non-ultra* to her endevours?"[68] The final paragraph in Power's *Experimental Philosophy in Three Books*... begins by brushing aside the Aristotelians. He then uses the clockwork analogy to demonstrate that experimental philosophers must look into the wheelwork of a watch and not judge merely by external appearances:

> This is the Age wherein (me-thinks) Philosophy comes in with a Spring-tide; and the Peripateticks may as well hope to stop the Current of the Tide, or (with *Xerxes*) to fetter the Ocean, as hinder the overflowing of free Philosophy: Methinks, I see how all the old Rubbish must be thrown away, and the rotten Buildings be overthrown, and carried away with so powerful an Inundation.... I think it is no Rhetorication to say, That all things are Artificial; for Nature it self is nothing else but the Art of God. Then, certainly, to find the various turnings, and mysterious process of this divine Art, in the management of this great Machine of the World, must needs be the proper Office of onely the Experimental and Mechanical Philosopher. For the old Dogmatists and Notional Speculators, that onely gaz'd at the visible effects and last Resultances of things, understood no more of Nature, than a rude Countrey-fellow does of the Internal Fabrick of a Watch, that onely sees the index and Horary Circle, and perchance hears the Clock and Alarum strke in it: But he that will give satisfactory Account of these Phaenomena, must be an Artificer indeed, and one well skill'd in the Wheel-work and Internal Contrivance of such Anatomical Engines.[69]

By Power's time, the mechanical metaphor was becoming a way of thought.

Boyle uses the analogy in a manner comparable to Glanvill and Power:

> he, that would thoroughly understand the nature of a watch, must not rest satisfied with knowing in general, that a man made it, and that he made it for such uses; but he must particularly know, of what materials the spring, the wheels,

> the string or chain, and the balance are made: he must know the number of the wheels, their bigness, their shape, their situation and connexion in the engine, and after what manner one part moves the other in the whole series of motions,... In short, the neglect of efficient causes would render physiology useless; but the studious indagation of them will not prejudice the contemplation of final causes....[70]

But elsewhere Boyle follows Descartes in using the analogy to demonstrate "That the hypotheses of philosophy only shew that an effect may be produced by such a cause, not that it must":

> he, that in a skilful watch-maker's shop shall observe how many several ways watches and clocks may be contrived, and yet all of them shew the same things; and shall consider how apt an ordinary man, that had never seen the inside but of one sort of watches, would be to think, that all these are contrived after the same manner, as that, whose fabrick he has already taken notice of; such a person, I say, will scarce be backward to think, that so admirable an engineer as nature, by many pieces of her workmanship, appears to be, can, by very various and differing contrivances, perform the same things; and that it is a very easy mistake for men to conclude, that because an effect may be produced by such determinate causes, it must be so, or actually is so.[71]

The passage which follows points out tactfully that Epicurus and some of his modern followers claim only that their hypotheses discover the "*possible* causes of the phaenomenon they endeavour to explain" (italics added). Boyle questions whether this is enough, and again supports his argument with a simile from the Strasbourg clock: "As it is one thing for a man ignorant of the mechanicks to make it plausible, that the motions of the famed clock at *Strasburg*, are performed by the means of certain wheels, springs, and weights, &c. and another to be able to describe distinctly the magnitude, figures, proportions, motions, and, in short, the whole contrivance either of that admirable engine, or some other capable to perform the same things."[72]

Boyle's writings are so extensive, his interests so wide, and his use of the clock analogy so frequent—we have earlier mentioned his self-conscious reference to "the *Strasbourg* clock I have so often alluded to"[73]—that he is not always completely consistent. In respect to the analogy with which we have been concerned, he does at one point seem to permit hypotheses about things we do not see. Boyle suggests that denying the possibility of extending mechanical principles from "natural bodies, whose bulk is manifest and their structure visible... to such portions of matter, whose parts and texture are invisible; may perhaps look to some, as if a man should allow, that the laws of mechanism may take place in a town clock, but cannot in a pocket-watch...."[74] This diligent experimental philosopher thinks in terms of the processes of investigation: "An ordinary watch-maker may be able to understand the curiosest contrivance of the skilfullest artificer, if this man take care to explain his engine to him...."[75]

Locke's use of the clock analogy to relate external appearance to internal mechanism invites comparison with Boyle, Power, and Descartes. The clock is the Strasbourg clock of Boyle and the "gazing countryman" is the "rude Countrey-fellow" of Power, but the essential concept goes back to Descartes. Under the important heading of *The Nominal and Essence Different*, Locke maintains that "had we such a knowledge of... man... as it is possible angels have, and as it is certain his Maker has, we should have a quite other idea of his essence than what now is contained in our definition of that species... as is his who knows all the springs and wheels and other contrivances within of the famous clock at Strasbourg, from that which a gazing countryman has for it, who barely sees the motion of the hand, and hears the clock strike, and observes only some of the outward appearances."[76]

In book 2 of his *Essay concerning Human Understanding*, Locke suggests that the ideas in the mind must not necessarily be assumed to be the physical objects that they represent. Here, surely, is the potential for idealism, though Locke does not take it as far as does Berkeley. Dr. Johnson's famous remark illustrates excellently the popular view of Berkeley's philosophy. He said to a gentleman who supported Berkeley's concepts: "Pray, Sir, don't leave us; for we may perhaps forget to think of you, and then you will cease to exist." One might think that the idealism of Berkeley would hardly lend itself to a mechanistic analogy. Indeed, in his later *Siris* (1744) he criticizes both Descartes and Leibniz, claiming that the latter's world went "like a clock or machine by itself, according to the laws of nature, without the immediate hand of the artist."[77] But significantly—in his *Principles of Human Knowledge* (1710), produced five years before the Leibniz-Clarke correspondence—Berkeley had found it necessary to state his position in mechanistic terms. He defends "idealism" by asking the rhetorical question, Why, if it is the spirit that "produces every effect," an artist should work on the movement of a watch, since "intelligence" can "create" the time on the face anyway? Similarly, why does God trouble to produce the infinitely more complex clockwork of nature:

> If it be a spirit that immediately produces every effect by a fiat, or act of his will, we must think all that's fine and artificial in the works, whether of man or nature, to be made in vain. By this doctrine, tho an artist has made the spring and wheels, and every movement of a watch, and adjusted them in such a manner, as he knew wou'd produce the motions he design'd; yet he must think all this done to no purpose, and that it is an intelligence which directs the index, and points to the hour of the day. If so, why may not the intelligence do it, without his being at the pains of making the movements, and putting them together? Why does not an empty case serve as well as another; and how comes it to pass, that whenever there is any fault in the going of a watch, there is some corresponding disorder to be found in the movements, which being mended by a skilful hand, all is right again? The like may be said of all the clockwork of nature, great part whereof is so wonderfully fine and subtile, as scarce to be discern'd by the best microscope.[78]

Berkeley's reply seems to give a new twist to the clock analogy that differentiates between external appearances and internal mechanism. According to him, the artisan frequently learns from nature about "framing artificial things, for the use and ornament of life." He is always obliged to work according "to the rules of mechanism," whereas for God this is not essential: "It cannot be denied that God . . . might . . . cause all the motions on the dial-plate of a watch, tho no body had ever made the movements... but yet if He will act agreeably to the rules of mechanism, by Him for wise ends establish'd... it is necessary that those actions of the watch-maker whereby he makes the movements... precede the production of the aforesaid motions, as also that any disorder in them be attained with the perception of some corresponding disorder, in the movements, which being once corrected all is right again." A miracle only occurs on those rare occasions when God chooses to interrupt the normal mechanical workings of his universe.[79] Naturally, a system based on the concept that "to be is to be perceived" is not sympathetic to mechanistically oriented philosophers. Berkeley achieved little prominence until criticized in Thomas Reid's *Inquiry into the Human Mind* (1764).

The biological scientists would later also question the value of the clock model because it could not satisfactorily illustrate organic growth. But John Ray (1627–1705), whose plant classifications greatly influenced botany, felt the need (like Berkeley) to use a mechanistic idiom during the horological revolution. In Ray's work, *The Wisdom of God...*, he too uses the clockwork analogy. His purpose is to demonstrate that "the infinitely wise Creator hath shewn in many Instances, that he is not confin'd to one only Instrument for the working one Effect, but can perform the same thing by divers Means." Just as "Clocks or other Engines" can be moved "by Springs instead of Weights.... So, tho Feathers seem necessary for flying, yet hath he enabled several Creatures to fly without them."[80]

Perhaps the final word from the philosophers using this particular analogy should be left to Cotes, in his preface to the second edition of the translation of Newton's *Principia*. In an analogy that we have traced thus far from Descartes, the reference to the "vortices" should not go unnoticed:

> It is reasonable enough to suppose that from several causes, somewhat differing from one another, the same effect may arise; but the true cause will be that from which it truly and actually does arise; the others have no place in true philosophy. The same motion of the hour-hand in a clock may be occasioned either by a weight hung, or a spring shut up

> within. But if a certain clock should be really moved with a weight, we should laugh at a man that would suppose it moved by a spring, and from that principle, suddenly taken up without further examination, should go about to explain the motion of the index; for certainly the way [he] ought to have taken would have been actually to look into the inward parts of the machine, that he might find the true principle of the proposed motion. The like judgment ought to be made of those philosophers who will have the heavens to be filled with a most subtile matter which is continually carried round in vortices. For if they could explain the phenomena ever so accurately by their hypotheses, we could not yet say that they have discovered true philosophy and the true causes of the celestial motions, unless they could either demonstrate that those causes do actually exist, or at least that no others do exist.[81]

Unlike the previously mentioned cluster of analogies—in which Donne and Herbert preceded Descartes and Clauberg—the philosophers come first in the pattern with which we are here concerned. The clock analogy for illustrating the difference between external appearance and internal mechanism does not seem to have been employed by literary writers until the middle of the eighteenth century. At that time, Richardson and Johnson both used it to indicate the difference between two types of authors.

In a letter to Sarah Fielding of 1756, Richardson praises her writing at the expense of her brother's: "What a knowledge of the human heart! Well might a critical judge of writing say, as he did to me, that your late brother's knowledge of it was not (fine writer as he was) comparable with your's. His was but as the knowledge of the outside of a clock-work machine, while your's was that of the finer springs and movements of the inside."[82] In Boswell's *Life*, Johnson implies much the same comparison between Richardson and Henry Fielding: "Sir, there is more knowledge of the heart in one letter of Richardson's, than in all 'Tom Jones.'"[83] Four years earlier (1768), Johnson actually used Richardson's clock analogy for demonstrating the difference between that author and Henry Fielding: "There was as great a difference between them as between a man who knew how a watch was made, and a man who could tell the hour by looking on the dial-plate."[84]

Essentially, the relatively simple analogy that we have been tracing through from Descartes to Dr. Johnson illustrates that a knowledge of external appearance does not ensure a knowledge of internal mechanism. But the examples indicate how the authors were able to adapt the analogy to their own particular interests.

Extended Clock Metaphors Used by Simon Patrick, Hobbes, and Shaftesbury

We shall now consider how thinkers as different as Simon Patrick, Hobbes, and Shaftesbury used more extended clock metaphors during the horological revolution for the purpose of emphasizing a particular argument. The extended metaphor is used by Patrick to promote experimental philosophy, by Hobbes to examine the body politic, and by Shaftesbury to discuss the concept of beauty in the nature of society.

Patrick was one of the Cambridge Platonists who seems to have fallen into some disfavor at the Restoration for Puritan sympathies; this appears to have cost him the mastership of Queen's College. Patrick's fascinating defense of the "Latitude-Men" (1662) uses the story of a broken old clock in a farmer's field to demonstrate the ineffectiveness of Aristotelian and "Scholastick Philosophy," as represented by a "Peripatetick artificer," a locksmith "well read in *Clock-Philosophy*," and the "Farmer's Son . . . newly come from the University." By comparison, the practical "Landlord of this Farmer," who represents the "atomicall" philosophy, "had used to take in pieces his own Watch and set it together again," On the basis of experience, he explains how the clock works and what is needed to repair it. Patrick uses the parable of the clock to demonstrate that

> it must be the Office of Philosophy to find out the process of this Divine Art in the great automaton of the world, by

> observing how one part moves another, and how those motions are varied by the severall magnitudes, figures, positions of each part, from the first springs or plummets, as I may say, to the hand that points out the visible and last effects.[85]

At the beginning of his major work, Hobbes introduced "the Nature of this Artificiall man," the Leviathan, through an analogy with automata and watches. The clock as an analogy for the mechanical nature of the material man was by then well understood. Hobbes could also therefore simultaneously use the extended metaphor to illustrate the parallel between the individual man and the state as a body corporate that is essential to his work:

> Nature (the Art whereby God hath made and governes the World) is by the *Art* of Man, as in many other things, so in this also imitated, that it can make an Artificial Animal. For seeing life is but a motion of Limbs, the beginning whereof is in some principall part within; why may we not say, that all *Automata* (Engines that move themselves by springs and wheeles as doth a watch) have an artificial life: For what is the *Heart*, but a *Spring*; and the *Nerves*, but so many *Strings*; and the *Joynts*, but so many *Wheeles*, giving motion to the whole Body, such as was intended by the Artificer? *Art* goes yet further, imitating that Rationall and most excellent worke of Nature, *Man*. For by Art is created that great LEVIATHAN called a COMMON-WEALTH, or STATE, (in latine CIVITAS) which is but an Artificiall Man; though of greater stature and strength than the Naturall, for whose protection and defence it was intended; and in which, the *Soveraignty* is an Artificiall *Soul*, as giving life and motion to the whole body. . . .

Hobbes then lists the contents of the four parts of *Leviathan*. Their relationship to the extended metaphor of the watch demonstrates its importance:

> To describe the Nature of this Artificiall man, I will consider

> First, the *Matter* thereof, and the Artificer; both of which is *Man*.
> Secondly, *How*, and by what *Covenants* it is made; what are the *Rights* and just *Power* or *Authority* of a *Soveraigne*; and what it is that *preserveth* and *dissolveth* it.
> Thirdly, what is a *Christian Common-wealth*.
> Lastly, what is the *Kingdome of Darkness*.[86]

Man, for Hobbes, would seem to be an automaton whose thought, actions, and art are all mechanically determined. Of what purpose then the exhortation to use one's free will and "Authorize and give up my Right of Governing my selfe" in order to obtain security from a Leviathan? It is an irony that one of the greatest philosophers should have such a flaw in his philosophy.[87]

Unlike Patrick and Hobbes, Shaftesbury subscribes to the so-called "eighteenth-century optimism" associated with Leibniz and Locke. Indeed, Locke was not only Shaftesbury's mentor, but had brought him into the world. In the age of Boyle and Newton—when Spinoza was a lens grinder, Leibniz invented a calculating machine, and Locke was a physician—there is a danger in differentiating too precisely between the scientist (or even technician) and the philosopher. The extent to which such men were in contact with one another should also not be overlooked. Locke, for example, apart from his connection with Boyle, resided with Cudworth's daughter, and was consulted by Newton on the question of coinage.

Shaftesbury's optimism about human nature obliges him to react to the mechanistic philosophy of Hobbes, though he does not directly name him. Ironically, Shaftesbury, too, feels obliged to use the idiom of the watch: "Modern Projectors, I know, wou'd willingly rid their hands of these *natural* Materials; and wou'd fain build after a more uniform way. They wou'd new-frame the Human Heart; and have a mighty fancy to reduce all its Motions, Ballances and Weights to that one Principle and Foundation of a cool and deliberate Selfishness."[88]

It might be thought that Shaftesbury is here using the horological idiom merely because it suits the subject of the mechanistic philosophy. But this is not so. It is precisely when he is making a case for the inherent goodness of human nature that Shaftesbury employs his most extended metaphor of the clock:

> You have heard it (my Friend!) as a common Saying, that *Interest governs the World*. But, I believe, whoever looks narrowly into the Affiars of it, will find that Passion, Humour, Caprice, Zeal, Faction and a thousand other Springs, which are counter to Self-Interest have as considerable a Part in the Movements of this Machine. There are more Wheels and Counter-Poises in this Engine than are easily imagin'd. Tis of too complex a kind to fall under one simple view, or be explain'd thus briefly in a word or two. The Studiers of this Mechanism must have a very partial Eye, to overlook all other Motions besides those of the lowest and narrowest compass. Tis Hard, that in the Plan or Description of this Clock work, no Wheel or Ballance shou'd be allow'd on the side of the better and more enlarg'd Affections; that nothing shou'd be understood to be done in Kindness or Generosity; nothing in pure Good-Nature or Friendship, or thro any social or natural Affection of any kind: when, perhaps, the main Springs of this Machine will be found to be either these very natural Affections themselves, or a compound kind deriv'd from them, and retaining more than one half of their Nature.[89]

Analogies That Differentiate between Clocks and Watches

In conclusion, let us consider the philosopher's use of the clock analogy that points to the difference in the size and the precision of clocks and watches. The first example of this that I have discovered appears in Charleton's *Physiologia Epicuro-Gassendo-Charltonia* (1654), the work that introduced into England the suitably modified atomic theories of Epicurus. As occurs with so many of the clock analogies used by philosophers, it is employed to explain one of the most important concepts in the writer's system. To illustrate the atom, Charleton needs an analogy for the minute "Mechanicks of Nature." In an age when precision engineering was virtually unknown, the "exquisite Artist" who could make a complicated watch small enough to be "set . . . upon a ring," provided an excellent analogy for the very fine mechanism of God's handiwork:

> Consider we first, that an exquisite Artist will make the movement of a Watch, indicating the minute of the hour, the hour of the day, the day of the week, moneth, year, together with the age of the Moon, and time of the Seas reciprocation; and all this in so small a compass, as to be decently worn in the pall of a ring: while a bungling Smith can hardly bring down the model of his grosser wheels and balance so low, as freely to perform their motions in the hollow of a Tower. If so; well may we allow the finer fingers of that grand Examplar to all Artificers, Nature, to distinguish a greater multiplicity of parts in one Grain of *Millet feed*, then ruder man can in that great Mountain, Caucasus; nay, in the whole *Terrestrial Globe.*[90]

Ring watches are still considered something of a mechanical novelty, but for a proper appreciation of the analogy one must think of it within the context of its time. Just four years before Charleton's work, a commentary on the Pentateuch glossing Exodus 35:32 helps us to evaluate the analogy of the ring watch in contemporary terms. On the Bible's "to devise curious works," the commentary says that "a certain artificer set a watch-clock upon a ring that Charls the Fifth wore upon his finger. King *Ferdinand* sent to *Solyman* the Turk, for a present, a wonderful globe of silver, of most rare and curious device; daily expressing the hourly passing of the Time, the motions of the Planets, the change and full of the Moon; lively expressing the wonderful conversions of the Celestial frame."[91]

Seven years later, when the British horological revolution was just beginning, the author of the *History of Most Manual Arts* is proud of man's newly acquired ability to make very small mechanisms: "The wit of man hath been luxuriant and wanton in the Inventions of late years; some have made Watches so small and light, that Ladies hang them at their ears like pendants and jewels; the smallness and variety of the tools that are used about these small Engines, seem to me no less admirable then the Engines themselves." The author concludes by comparing, like

Charleton, the relative skills required for making watches and clocks: "There is more Art and Dexterity in placing so many Wheels and Axles in so small a compass (for some French Watches do not exceed the compass of a farthing) then in making Clocks and greater Machines."[92]

Boyle is naturally aware of the particularly fine mechanism of God's work: "The structure even of the rarest watch is incomparably inferior to that of the human body." Although he feels one should not claim "that the laws of mechanism may take place in a town clock, but cannot in a pocket-watch,"[93] he makes a clear distinction elsewhere between a clockmaker and a watchmaker. Boyle feels that "God's wisdom is recommended as well by the variety, and consequently the number of the kinds of living creatures, as by the fabrick of each of them in particular. For the skill of human architects and other artists is very narrow, and for the most part limited to one or a few sorts of contrivements. Thus many an architect can build a house well, that cannot build a ship; and (as we daily see) a man may be an excellent clockmaker, that could not make a good watch."[94] Once the more accurate watches came into production after about 1675, both their reputation and the demand for them grew rapidly. Tompion, though the son of a blacksmith, was himself hardly the "bungling Smith" of Charleton's analogy.

This sentiment about the differences in the standards of engineering required in making a watch and a clock (particularly a large turret clock) adds piquancy to one of Boswell's reflections on a remark of Johnson. The last example in one of our previous patterns of analogies was Dr. Johnson's famous statement regarding Richardson and Fielding, "That there was as great a difference between them as between a man who knew how a watch was made, and a man who could tell the hour by looking on the dial-plate." Boswell comments on this (and his evaluation is perhaps closer to our own): "But I cannot help being of opinion, that the neat watches of Fielding are as well constructed as the large clocks of Richardson, and that his dial-plates are brighter,"[95]

As the period progressed from the time of Descartes on into the horological revolution, the greater accuracy and popularity of watches, as well as the greater skill frequently required in making them, resulted in watches slowly taking over from clocks the role of mechanical metaphor. But there is no clearly discernible transition, and I have therefore generally considered the terms *watch* and *clock* as interchangeable in describing the metaphor. However, when the time came for God to be revealed in his new role of mechanical engineer, the mechanically oriented Western civilization naturally worshipped him as a divine watchmaker rather than as a "bungling Smith."

The Watchmaker God

Leibniz, Clarke, and Newton

A watchmaker God implies a universe that is rational, orderly, and comprehensible. His mechanism invited increasingly penetrating investigation by the experimental philosophers. During the course of these investigations—and perhaps not unconnected with the concurrent rise of technology in science, Whiggism in politics, and sentiment in literature—philosophers came to the optimistic conclusion that God had constructed the best of all possible worlds for men. Man's life, which Hobbes had claimed to be basically "nasty, brutish and short," was now declared inherently good.

The argument—with which we are familiar from the first epistle of Pope's *Eassy on Man*—is that all our senses have been nicely adjusted by God to the needs which we have. A "microscopie eye" or an overacute sense of touch, smell, or hearing would make life intolerable. (The concept that such adjustment came through development and natural selection would seem to be unsympathetic to the mechanical model.) The optimism that Pope's *Essay* implies is generally associated with Locke and Leibniz, but the same argument is earlier made at length in Henry More's *Antidote against Atheism.* It is also to be found in the *De Homine* (1677) of Matthew Hale, a former chief justice of England, who describes "the admirable accommodation of

Sensible Faculty to the Objects of Sense, and of those Objects to it, and to both of the well-being of the Sensible Nature."[1]

Locke's range of examples in his *Essay concerning Human Understanding* is much the same as that of those Pope would later use. But in closing, he feels the need to restrain his imagination; he brings his argument (so important for eighteenth-century optimism) back to the concrete—to the market, the exchange, and, above all, the clock:

> And if by the help of such microscopical eyes (if I may so call them) a man could penetrate further than ordinary into the secret composition and radical texture of bodies, he would not make any great advantage by the change, if such an acute sight would not serve to conduct him to the market and exchange.... He that was sharp-sighted enough to see the configuration of the minute particles of the spring of a clock, and observe upon what peculiar structure and impulse its elastic motion depends, would no doubt discover something very admirable: but if eyes so framed could not view at once the hand, and the characters of the hour-plate, and thereby at a distance see what o'clock it was, their owner could not be much benefitted by that acuteness; which, whilst it discovered the secret contrivance of the parts of the machine, made him lose its use.[2]

The question of perspective—not only in terms of size but in the use of the senses and the mind—is also a theme in Swift's *Gulliver's Travels.*

Leibniz, even more than Locke, championed eighteenth-century optimism, maintaining that man lives in the best of all possible worlds with the optimum of order and harmony. It was Liebniz's *Essais de théodicée*—and his problems of reconciling evil and free will with the concept of preestablished harmony—that Voltaire would attack in *Candide* (1759). It is fundamental to Leibniz's monadology that the "mutual connection or accommodation of all created things to each other ... is the means of obtaining the greatest variety possible, but with the greatest possible order; that is to say, this is the means of attaining as much perfection as possible."[3]

Leibniz's controversy with Samuel Clarke is related to his view of preestablished harmony. It serves also as an admirable background to the argument from design which attempted to prove the existence of God through the analogy between his universe and a watch. The finer points of the controversy between Leibniz and Clarke do not concern us, what does is the manner in which both parties accepted the metaphor of watch and watchmaker for the universe and its God. It has been suggested that "Newton himself supplied Clarke with his critical questions and answers."[4] Caroline of Ansbach, the future queen of England, acted as mediator in this philosophical battle of "mighty opposites."

Leibniz held that in order to make phenomena intelligible "we must not unnecessarily resort to God."[5] Furthermore, once God had set in motion the infinitely complex clock of the universe, the preestablished harmony of that universe would preclude any theory that the clock might need correcting from time to time. Newton's *Opticks* would appear to suggest otherwise: "Some other Principle was necessary for putting Bodies into Motion, and now they are in Motion, some other Principle is necessary for conserving the Motion." This is because it is natural that "Motion is much more apt to be lost than got, and is always upon the Decay." Newton also speaks of "some inconsiderable Irregularities" in the heavenly bodies, "which may have arisen from the mutual Actions of Comets and Planets upon one another, and which will be apt to increase, till the System wants a Reformation."[6]

Newton's clock needs winding up from time to time; Leibniz's does not. But, as frequently occurs in disputes between academics, the impact of the mundane undoubtedly made its contribution. Leibniz was a Trinitarian and Newton sufficiently anti-Trinitarian to require a special dispensation for holding his professorship. Worse still, Newton claimed that Leibniz had plagiarized his invention of the differential calculus, a claim that is today not considered valid.

Like Newton's, Clarke's orthodoxy was not beyond suspicion; *The Scripture Doctrine of the Trinity* (1712) had probably cost him his preferment in the church. Voltaire, who admired Clarke, tells the story of how the bishop of Lincoln prevented his elevation. He informed Princess Caroline that though Clarke was the most learned and honest man in her dominions, he had but one defect—he was not a Christian. Whatever the merit of this story, the letters are naturally written in a very different vein. Despite the argument and the animus, both parties assume that the mathematical and mechanical view of the world is almost as naturally accepted as is the existence of God.

Leibniz's opening letter sets the tone:

> Sir Isaac Newton, and his followers, have also a very odd opinion concerning the work of God. According to their doctrine, God Almighty wants to wind up his watch from time to time: otherwise it would cease to move. He had not, it seems sufficient foresight to make it a perpetual motion. Nay, the machine of God's making, is so imperfect, according to these gentlemen; that he is obliged to clean it now and then by an extraordinary concourse, and even to mend it, as a clockmaker mends his work; who must consequently be so much the more unskilful a workman, as he is oftener obliged to mend his work and to set it right. According to my opinion, the same force and vigour remains always in the world, and only passes from one part of matter to another, agreeably to the laws of nature, and the beautifully pre-established order.[7]

Dr. Clarke points out, in his reply, that the concept of a clock that will continue of its own accord is one based purely on human values: "The reason why, among men, an artificer is justly esteemed so much the more skilful, as the machine of his composing will continue longer to move regularly without any further imposition of the workman is because the skill of all human artificers consists only in composing, adjusting or putting together certain movements, the principles of whose motion are altogether independent upon the artificer: such as are weights and springs, and the like; whose forces are not made, but only adjusted by the workman." Clarke insists that God does not need this type of credit. It is rather to his greater glory that "nothing is done without his continual government and inspection": "The notion of the world's being a great machine, going on without interposition of God, as a clock continues to go without the assistance of a clockmaker; is the notion of materialism and fate, and tends, (under pretence of making God a *supra-mundane intelligence*) to exclude providence and God's government in reality out of the world."[8]

Clarke's accusation regarding "materialism and fate" suggests Hobbesian atheism and predestination. In reply, Leibniz becomes more circumspect, but he keeps the same terms of reference: "I do not say, the material world is a machine, or watch, that goes without God's interposition; and I have sufficiently insisted, that the creation wants to be continually influenc'd by its creator. But I maintain it to be a watch that goes without wanting to be mended by him: otherwise we must say, that God bethinks himself again. No; God has forseen every thing beforehand; there is in his works a harmony, a beauty, already preestablished."[9]

Leibniz points out further that God's excellence arises from his wisdom, "whereby his machine lasts longer, and moves more regularly, than those of any other artist whatsoever." Clarke picks this up in his second reply, where he claims that the "wisdom of God appears, not in making nature (as an artificer makes a clock) capable of going on without him: (for that's impossible; there being no powers of nature independent upon God, as the powers of weights and springs are independent upon men;) but the wisdom of God consists in framing originally the perfect and complete idea of a work, which begun and continues.... by the continual uninterrupted exercise of his power and government."[10]

The way that they employ the clock metaphor is symptomatic of the political and theological background of both writers. Clarke, like Leibniz, is careful to avoid any implication that the Cartesian concept of the beast-machine be extended to cover the soul as well as the body of man. When Clarke says, "If the word *natural forces*, means here mechanical; then all animals and even men, are as mere machines as a clock," the context makes it evident that he does not intend men to be considered as machines. Elsewhere, he says that "to suppose, that all the motions of our bodies are necessary, and caused entirely... by mere mechanical impulses of matter... tends tends to make men be thought as mere machines, as Descartes imagined beasts to be...."[11]

Leibniz, too, wants to dissociate himself from any such concept as *l'homme-machine*. He maintains that a man's "body is truly a machine, acting only mechanically; and yet his soul is a free cause." The problem of trying to synchronize the action of body and soul is dealt with as follows: "'Tis true, that according to me, the soul does not disturb the laws of the body, nor the body those of the soul; and that the soul and body do only agree together; the one acting freely, according to the rules of final causes; and the other acting mechanically, according to the laws of efficient causes. . . . because God, foreseeing what the free cause would do, did from the beginning regulate the machine in such manner, that it cannot fail to agree with that free cause."[12] Of three possible arrangements for harmonizing soul and body (which he equates with two clocks) Leibniz proudly insists on the third: "To make the two clocks [*pendules*] at first with such skill and accuracy that we can be sure that they will always afterwards keep time together. This is the way of pre-established agreement [*consentement*]."[13]

The dualism of body and spirit had been inherited by the mechanistic philosophy and posed some real problems for the horological metaphor. Leibniz, as we have just seen, makes the spirit, through the operation of a superior cause, work in harmony with a mechanistic body. Descartes had earlier suggested that the pineal gland acted as a form of clearing house between the soul and a mechanical body. Others like Spinoza and Hobbes claimed to retain a belief in God while giving mechanical attributes to the soul. Hobbes, an important exponent of the mechanistic philosophy in England, claimed that the soul itself was material; Spinoza held the view that "the soul acts according to fixed laws, and is as it were an immaterial automaton." The Malebranchian occasionalists and the Cambridge Platonists attempted to deal with the problem by accepting, in varying degrees, the mechanistic idiom for the flesh, while giving greater stress to the importance of the spirit.

The Argument from Design

In Leibniz, Clarke, and Newton, the concept of a watchmaker God is implicit rather than explicit. But the argument from design—proving the existence of God through an analogy of his universe with a watch—was actively disseminated throughout the British horological revolution of 1660–1760, and, in the hands of theologians, long thereafter. It is true that though Cartesian philosophers related animals to clocks they had problems with the souls of men, while poets—who frequently related men to clocks, as we shall later see—were virtually unanimous in attacking the philosophers' concept of mechanical animals. During the horological revolution, however, all parties could agree that the heliocentric universe demonstrated the order and harmony for which man's most appropriate analogy was the watch.

The argument from design proceeds from the following analogy: the universe is like a macrocosm of which the watch is a microcosm. Once this is accepted—as it frequently was—the argument from design follows: the universe is like a watch; a watch demonstrates the existence of a watchmaker; therefore the universe demonstrates the existence of its own Master Craftsman. In a society where men of genius like Leibniz and Newton readily compared the universe with a watch, men of lesser intellect tended to forget that this was only an analogy. The more one overlooks the analogy and the more one believes that the universe really is an enormous mechanism, the stronger the argument from design becomes. Ultimately it possesses almost the full compulsion of a syllogism. If A is a universe, B is a watch, and C is a watchmaker:

> A is a macrocosm of B;
> B is made by C;
> Therefore A is made by a macrocosm of C (a divine Watchmaker

In an age which felt that the existence of God could be proved, the teleological argument for God's existence as the creator of an ordered universe was particularly attractive. Understandably, the use of this argument coincides closely, at least at first, with the expanding influence of clocks on the nature of society, particularly during the horological revolution of 1660–1760.

The argument from design is an aspect of Englightenment

history of which most readers will be aware. It is mentioned briefly in many history books, while Stephen's *History of English Thought in the Eighteenth Century* and Hallam's *Literature of Europe* give a few examples.[14] Palmer puts it this way: "The great symbol of the Christian God was the Cross, on which a divine being had suffered in human form. The symbol which occurred to people of scientific view was the Watchmaker. The intricacies of the physical universe were compared to the intricacies of a watch, and it was argued that just as a watch could not exist without a watchmaker, so the universe as discovered by Newton could not exist without a God who created it and set it moving by its mathematical law."[15]

The earliest reference to this argument that I have found in English literature occurs in a translation from the French of Philip Mornay's *A Worke concerning the Trewnesse of the Christian Religion* (1587). This is about the time that watches first began to be made in England. There are, however, far more examples in the period with which we are involved. Among them are the watch analogies that occur in John Smith's *Select Discourses* (1660), Lord Herbert of Cherbury's *De Religione Gentilium* (1663), John Spencer's *A Discourse concerning Prodigies* (1663), and N. Fairfax's *A Treatise of the Bulk and Selvedge of the World* (1674). Most of these references are short, though Fairfax continues in his somewhat facetious style for about two pages. He sees "the Watch-wrights craft [as] not only the Ape of Nature, but the very Tool, still in her hand."[16]

Matthew Hale, in *The Primitive Origination of Mankind* (1677), uses the fiction of a clock found in a field, much like the one in Simon Patrick's defense of the "Latitude-Men" (1662), to which reference has already been made. Like Patrick, he shows at some length how "Scholars of the several Schools" react to this. Hale's third and fourth scholars demonstrate an understanding of the argument from design. But "the Artist that made this Engin" confutes them all by a practical explanation of how a watch is made. Hale's conclusion is that "the plain, but Divine Narrative by the hand of *Moses*" is much more reliable than "the Hypotheses of the Learned Philosophers."[17] The fiction of a watch discovered in the open air by people who do not understand it is part of the fiction of the argument from design. We shall meet it again in the extended arguments of Nieuwentyt's *Religious Philosopher* (1718) and Paley's *Natural Theology* (1802). The half brother of this fiction (to be dealt with in a later chapter) is the reaction of primitive natives who think that a watch is a living animal. (Swift deals with the topical concept ironically when the king of Brobdingnag considers Gulliver to be a clockwork automation.)

Not all arguments from design are formally expounded. Boyle (in a manuscript at the Royal Society) asks, "And shall we readily allow so much foresight & contrivance to a Mechanicall artificer, and shall we scruple to allow much better mechanismes to (the Author even of Artificers) the Omniscient God himself, in the Production of his Great *Automaton*, the World?"[18] There are comparable references in Cudworth's *Treatise concerning Eternal and Immutable Morality* (published after his death);[19] Blackmore's *Creation* (1712);[20] and even Voltaire's epigram,

> Le monde m'embarasse, et je ne puis pas songer
> Que cette horloge existe et n'a pas d'Horloger.

John Ray's *Wisdom of God*... couples the argument from design with "Rule, Order, and Constancy" in the heavens. He quotes from Cicero, "Shall we... when we see an Artificial Engine, as a Sphere or Dyal, or the like, at first Sight acknowledge, that it is a Work of Reason and Art?" This and other classical sources have been suggested by others. Hallam indicates that the argument from design may go back to the disputed passage in Cicero's *De Natura Deorum* 2.34; Macaulay (in *Essay: Von Ranke*) compares Paley's use of the watch to Socrates' argument employing the statues of Polycletus and the pictures of Zeuxis against the atheism of Aristodemus.

Derham's *Astro-Theology: Or a Demonstration of the Being and Attributes of God from a Survey of the Heavens* (1715) is by the rector of Upminster and fellow of the Royal Society who, in 1696, published the first creditable manual of horology. Stephen feels that Derham's was still "the most popular version" of the argument when Paley wrote his *Natural Theology*. In his second chapter entitled "The great *REGULARITY* of the Motions of every Globe," Derham brings together the optimism and the order: "If we consider that those Motions are wisely ordered and appointed, being as various, and as regular and every way nicely accomplished, as the World and its Inhabitants have occasion for. This is a manifest sign of a wise and kind, as well as omnipotent CREATOR and ORDERER of the World's affairs, as that of a Clock, or other Machine is of Man."[21] The Words *regularity* and *orderer* are emphasized by the largest print available; they underline the widely accepted association of the clock with those qualities.

In *The Light of Nature Pursued* (1768), Tucker has a good deal to say about the clock universe and the "divine Artist" who directs it. He discusses both the Newtonian and Leibnizian view of that universe, though without directly alluding to the philosophers.[22] Tucker has been considered the immediate influence on Paley's *Natural Theology*, but Chamberlayne's translation of Nieuwentyt's *Religious Philosopher* (1718) warrants direct comparison. The first of the following passages is from Nieuwentyt and the second from Paley:

> ... and should farther observe, that those Wheels are made of *Brass*, in order to keep them from Rust; that the Spring is of *Steel*, no other Metal being so proper for that Purpose; that over the Hand there is placed a clear Glass; in the Space of which, if there were any other but a transparent Matter, he must be at the Pains of opening it every time to look upon the Hand. ...
>
> We take notice that the wheels are made of brass, in order to keep them from rust; the springs of steel, no other metal being so elastic; that over the face of the watch there is placed a glass, a material employed in no other part of the work, but in the room of which, if there had been any other than a transparent substance, the hour could not be seen without opening the case.[23]

William Paley, archdeacon of Carlisle, made a remarkably good living and reputation as an apologist for the Christian religion. As a measure of his fame, he receives six double-column pages (more than David Hartley or even Robert Boyle) in the *Dictionary of National Biography*, but in the last decade he seems to have been dropped from the *Britannica*. Paley's *Principles of Moral and Political Philosophy* (1785) went through fifteen editions in his lifetime, and *A View of the Evidences of Christianity* (1794) went through fifteen editions in seventeen years, but the *Natural Theology, or Evidence of the Existence and Attributes of the Deity collected from the Appearances of Nature* (1802) did best of all. The twentieth edition had already appeared by 1820. In 1885 Darwin's theory was beginning to shake the foundations of religion from the direction of the plant analogy rather than the mechanical one with which Paley is particularly concerned. The *Natural Theology*—at that time over four score years of age and still with an active following—was suitably modified in order to meet the new conditions. There was published an edition by F. Le Gros Clark "revised to harmonize with modern science."[24]

The basis of Paley's *Natural Theology* is the proof of the existence of God by the argument from design. The first two chapters ("State of the Argument," and "State of the Argument Continued") and part of the third deal exclusively with what was known as "Paley's watch"; the whole of the work (almost six hundred pages) is constructed around the "Argument."

Paley begins by asking rhetorically why a watch is different from a stone. His reply shows that he is proud of being able to describe a watch. One hundred and fifty years had made a big difference in the general level of technical description. We, of course, known where Paley's argument will lead: "The watch must have had a maker; that there must have existed, at some time and at some place or another, an artificer or artificers who formed it for the purpose we find it actually to answer; who comprehended its construction, and designed its use."[25] Paley maintains that this conclusion would not have been weakened even if assailed by some eight possible objections into which he then goes at length. These are reminiscent of the points of view of such thinkers as Newton, Democritus, and Berkeley.

Chapter 2 opens with an argument that bears interesting comparison with Fontenelle's earlier objection to Cartesian mechanism. This was the suggestion that, unlike dogs, two watches were not able to create a third. Inventions surely have a direct relationship to the state both of man's dreams and of his technology. In the eighteenth century, men dreamt and wrote frequently of going to the moon, yet the dream could not be realized until it was combined with twentieth-century technology. But the advance of technology would also appear to provide a feedback for the dreams. One hundred years or so after Fontenelle, Paley no longer laughs at the possibility of the self-perpetuating potential of automatons:

> Suppose, in the next place, that the person who found the watch, should after some time, discover, that in addition to all the properties which he had hitherto observed in it, it possessed the unexpected property of producing in the course of its movement another watch like itself; (the thing is conceivable;) that it contained within its mechanism, a system of parts, a mould for instance, or a complex adjustment of laths, files, and other tools, evidently and separately calculated for this purpose; let us enquire, what effect ought such a discovery to have upon the former conclusion? [26]

Though the language is lucid enough, the argument leads tortuously onward, and 450 pages later little has changed. Ultimately, Paley states concisely what he has stated previously at no mean length: "Upon the whole; after all the schemes and struggles of a reluctant philosophy the necessary resort is to Deity. The marks of design are too strong to be got over. Design must have a designer. That designer must have been a person. That person is God." [27]

Mechanistic versus Organic Models in the Century before Darwin

The impact of Paley's argument from design during a considerable part of the nineteenth century might lead one to the conclusion that the watch analogy retained the same influence that it achieved during the horological revolution. This is demonstrably untrue. During the eighteenth century, a clearer distinction developed between professions. The clock analogy—at first so radical and so hard to reconcile with orthodox views—

remained respectably employed by theologians in the cause of piety. But philosophers and scientists became more and more disenchanted with the mechanistic model. (So did the Romantic poets and, above all, the dark Romantics like Hoffmann, Poe, and Baudelaire.)

Hume discusses the dichotomy in his *Dialogues concerning Natural Religion* (1779). The main dialogue is between Cleanthes, the theist, and Philo, the sceptic, who is closer to the author. In a work of twelve parts, Cleanthes (in part 2) sets up the standard argument from design: the world is "nothing but one great machine . . . we are led to infer by all the rules of analogy that the causes also resemble; and that the Author of nature is somewhat similar to the mind of man."[28] At the center of the work (in parts 6 and 7), Philo establishes that "the world plainly resembles more an animal or a vegetable, than it does a watch or a knitting loom." Therefore, its cause ought rather to be ascribed to generation or vegetation than to reason or design."[29]

There is a great deal in the *Dialogues* on the subject of watch and plant analogies,[30] and the concentration of this material in the structurally significant beginning, middle, and end of the work stresses the importance that it has in Hume's argument. It is true that Philo may wish to question the use of analogies altogether, and thereby question the process by which men frequently explain the "idea" of a God. What concerns us, however, is the way in which the developing climate of opinion became more receptive to biological and, in particular, botanical analogies. The change coincides with a new trend in scientific investigation that leads ultimately to Darwin's *Origin of Species*. But it is also observable in the poetic inspiration Goethe derived from extensive biological experiments or the ability of Shelley to indentify himself with a "sensitive plant."

There are implicit questionings of the mechanical model from the beginning of the horological revolution. No analogy can ever be completely satisfactory. For a time, the discrepancies were generally overlooked because the model was topical, readily understood, and suited to explaining difficult concepts in concrete terms. The main problem with explaining organic life through a mechanical model lies in the growth and immeasurably greater complexity of the former. During the seventeenth century, philosophers questioned the place of the soul and the rational mind in the mechanical model. But apart from this its appeal was so conformable to the *Zeitgeist* that in other respects they tended to accept it.

However, one can detect the direction from which difficulties would arise. Boyle suggests that the animal is a "total machine," each part of which is "a subordinate engine";[31] Locke awkwardly compares a watch whose parts are repaired with a tree or horse that in growth or decline retains its identity;[32] and Leibniz feels the need to insist that "a natural machine remains a machine even in its smallest parts."[33]

Leibniz's stress on the organic quality of bodies is not Cartesian. In fact, his reaching out for a vitalist rather than a mechanical model is reminiscent of Aristotle's words on the subject in *De Motu Animalium*:

> Animals have parts of a similar kind, their organs, the sinewy tendons to wit and the bones; the bones are like the wooden levers in the automaton, and the iron; the tendons are like the strings, for when these are tightened or released movement begins. However, in the automata and the toy wagon there is no change of quality, though if the inner wheels became smaller and greater by turns there would be the same circular movement set up. In the animal the same part has the power of becoming now larger and now smaller, and changing its form, as the parts increase by warmth and again contract by cold and change their quality.[34]

But Aristotle moves away from his very limited mechanical model, whereas Leibniz remains much influenced by the more sophisticated model of his age.

The stress that Leibniz does put on the superiority of organic mechanisms results both from his religious bent and from the recent discoveries (particularly of cells and spermatozoa) of Leeuwenhoek. Yet Leibniz still expresses himself in mechanistic terms: a "body is *organic* when it forms a kind of automaton or natural machine, which is a machine not only as a whole but also in its smallest observable parts."[35] This concept appears to run through the writings of Leibniz. In *Two Dialogues on Religion*,

Polidore is brought around to the author's point of view, and wonders "at the marvellous structure of organic bodies, the smallest part of which supasses in craftsmanship all the machines which man can invent."[36] Perhaps Leibniz's point is at its clearest in the *Monadology*. He argues there that the tooth of a brass wheel does not itself contain machines, whereas machines of nature have infinite complexity and "are still machines in their smallest parts."[37]

Despite Paley's suggestion that watches might give birth to watches, the upsurge in the biological sciences coincides with a reduced role for the clock analogy after about 1760. Before the earlier upsurge in Western technology, men had almost always used animate rather than inanimate models. (The atomistic theory is an important exception.) Just as the mechanical model continued and in newer forms (computers, cybernetics) is, in part, with us still, the earlier animate model carried over into the mechanistic period. Shakespeare can speak of cleansing "the foul body of the infected world" (*As You Like It* 2.7.60), and even Cudworth still calls it the "*Vast Automaton*, which some will have to be an Animal likewise."[38]

Gregory, in "The Animate and Mechanical Models of Reality," may be somewhat overstating an important case in saying, "When Descartes finally convinced the seventeenth century that physical nature was mechanical like a clock, and did not participate in animal qualities, an inveterate preconception received its *coup de grâce*." But Gregory shows how models can carry over from one period into another:

> In the later seventeenth century the famous philosopher John Locke still perpetuated an inveterate preconception. "All stones, metals, and minerals," he wrote, "are real vegetables; that is, grow organically from proper seeds, as well as plants." Nicholas Lemery still grumbled at chemists who searched through metals for their "seeds." Though chemistry was discarding this belief, and Locke himself implied that he had finally discarded it, the growth of metals from seeds had been an inveterate belief. Metals subject to sickness, ripened from seeds, and provided with souls, were inanimate substances conformed to an *animate*, and not to a *mechanical*, model.[39]

Gregory appears to see no clear trend in today's models: "Emergent doctrines have stiffened the adequacy of the mechanical model," but "many voices now protest that the animate model was too peremptorily dismissed."[40]

One of the advantages of history is that it permits us to see past trends in human affairs somewhat more clearly. When Kant, for example, rejects the watch analogy for organic life he does so partly because of its limitations in dealing with problems related to the newer biological science:

> one wheel in the watch does not produce the other, and, still less, does one watch produce other watches, by utilizing (organizing) foreign material; hence it does not of itself replace parts of which it has been deprived, nor, if these are absent in the original construction, does it make good the deficiency through the aid of the rest; nor does it, so to speak, repair itself if it goes out of order. But these are all things which we are justified in expecting from organized nature. An organized being, is, therefore, not a mere machine. For a machine has only moving power, whereas an organized being possesses inherent formative power... which cannot be explained by the capacity of movement alone, that is to say, by mechanism.[41]

However, McFarland believes that Kant was "unable to free himself from the watchmaker-watch analogy completely."[42]

It would be a mistake to think that the mechanistic model did not contribute to the theory of evolution. Hartley developed the mechanical theory of learning—through the association of ideas with pleasure and pain[43]—that has come down to us in the "behavioral" sciences. It is a curious point, frequently overlooked, that the "atheist" Hartley sought to relate the learning process in this life to its effect on the afterlife (*Observations* vol. 2). Erasmus Darwin ingeniously converts Hartley's learning process into a form of hereditary experience gained through association. Chapter 39 of *Zoonomia* (1794) is entitled "Generation," and is particularly revealing as an important milestone in the development of evolutionary ideas. Erasmus Darwin begins as follows:

> The ingenious Dr. Hartley in his work on man, and some other philosophers, have been of the opinion, that our immortal part acquires during this life certain habits of action or of sentiment, which become for ever indissoluble, continuing after death in a future state of existence; and add, that if these habits are of the malevolent kind, they must render the possessor miserable even in heaven. I would apply this igenious idea to the generation or production of the embryon, or new animal, which partakes so much of the form and propensities of the parent.
> Owing to the imperfection of language the offspring is termed a *new* animal, but it is in truth a branch or elongation of the parent; since a part of the embryon-animal is, or was, a part of the parent; and therefore in strict language it cannot be said to be entirely *new* at the time of its production; and therefore it may retain some of the habits of the parent-system.[44]

What was ultimately doomed was not so much the mechanistic metaphor as the hegemony of the clock metaphor. In McFarland's words, "It was not until the latter half of the eighteenth century that the notion that teleology in nature is analogous to the teleology of a mechanical artifact was seriously subjected to critical examination, first by Hume and later by Kant."[45] Although McFarland feels that Kant is unable "to free himself from the watchmaker-watch analogy completely," the passage previously quoted readily demonstrates that his view of the mechanical model is very different from Paley's. A more recent writer has said, "Paley's simile of the watch . . . must be replaced by the simile of the flower. The universe is not a machine but an organism with an indwelling principle of life. It was not made, but it has grown."[46]

During the second part of the eighteenth century, much of the value that the watch analogy had for poets, scientists, and philosophers was lost. This does not mean that the clock was not remembered—far from it. But when an old god falls there is always the risk that it will become the devil of future generations. A consideration of what clockwork and mechanized automatons came to mean to the Romantics will provide some interesting insights into their own literature as well as that of the preceding neoclassical period.

The first step is to consider the use of the clock metaphor by poets, and we shall find that Chaucer and Shakespeare, though intimately concerned with the question of time, are inhibited in varying degrees by the horological artifacts of their day. This limitation experienced by the two greatest poets of English medieval and Renaissance literature will emphasize the changes that take place in subsequent centuries.

Part III

The Influence on Literature

Poets and the Clock Metaphor

Early Horological References: Art as a Reflection of History

It is possible that clock metaphors antedate the weight-driven mechanical clock. We have already noted that terms like *horologium* and even *clock* derive from earlier timekeeping devices. All that we can say with certainty is that before the end of the fourteenth century clock metaphors and similes had already been occasionally used to explain animals, men, and the universe.

Saint Thomas Aquinas, in the course of denying free will to animals, compared the manner in which they function to "the movements of clocks and other works of human art." Aquinas adds: "Now artificial things are to human art as all natural things are to divine art."[1] The analogy in the *Summa theologica* occurs in the second half of the thirteenth century. What may be relevant is that Aquinas became the favorite pupil of Albertus Magnus, with whom he went to Cologne in 1248. Albertus was credited (reliably or otherwise) with the manufacture of automata, including a terra-cotta talking head.

Dante—who drew much of his theological learning from Aquinas, and especially from the *Summa*—appears to use a clock simile in the *Paradiso* to describe the workings of the universe. Interestingly, this comes immediately after a reference to Aquinas: "Then as the horologue, that calleth us, what hour the spouse of God riseth to sing her matins to her spouse that he may love her,/ Wherein one part drawing and thrusting other, giveth a chiming sound of so sweet note, that the well-ordered spirit with love swelleth;/ So did I see the glorious wheel revolve and render voice to voice in harmony and sweetness that may not be known except where joy maketh itself eternal." The references to Aquinas and Dante precede any mechanical clocks whose existence in known with certainty, and may refer to complex clepsydrae, or water clocks.[2]

In the latter part of the fourteenth century, Nicole Oresme used an unmistakable clock simile for the universe. At the request of Charles V, he translated Aristotle's *Ethics*, *Politics*, *Economics*, and *On the Heavens*. The last name appeared as *Le Livre du ciel et du monde* (1377). In a passage of this work, Oresme used the image of a clock to describe the workings of a universe motivated by God but moving under its own volition.[3]

As early as 1369, Froissart (1337–1404?) had written what is probably the longest poem about a clock; *Li Orloge Amoureus* runs to 1174 lines, and is an allegory relating the parts of a clock to a courtly lover overcome by the attractions of his lady. Froissart begins, "I can well compare myself to a clock. . . ." At four points he italicizes passages of several stanzas that deal respectively with the train and foliot, the dial (of which the daily movement, divided into "twenty-four hours," is exactly comparable to that which the pre-Copernican sun makes "around the earth in a natural day"), the striking part, and—"because neither a clock nor a poet/Go by itself"—the "orlogier," who is in daily attendance on the clock.[4] In 1493, Gaspari Visconti, in the preface to a sonnet, says that it "is put in the mouth of a lover who, looking at one of these clocks, compares himself to it."[5]

By 1400 then we have, on the continent, a few examples of clock analogies for the universe, for animals, and for men. Also, in Froissart's prosaic poetry we have a premonition that the clock and the Muses may not prove to be the best of bedfellows.

Chaucer (c. 1340–1400) travelled widely in France and Italy, and introduced many of the continental genres into the English tradition. But (with the exception of his *Treatise on the Astrolabe*) Chaucer's only direct allusion to a clock appears to be that referring to "a clokke or an abbey orlogge" in the *Nun's Priest's Tale*,[6] and to the cock as the "orloge" of small villages in *The*

Parliament of Fowles (l. 350). The relatively late appearance of mechanical clocks in England (Salisbury c. 1386 and Wells c. 1392) may be the cause. Certainly Chaucer's command of horology would not have been less than Froissart's.[7] His *Treatise on the Astrolabe* (c. 1391) is written for a remarkably erudite "Little Lewis my son"; Price has argued that Chaucer is also the author of a much more advanced work, *The Equatorie of the Planetis.*[8]

The *Astrolabe* has several references to clock time; eight mention "Equale houres," and eleven "inequale houres."[9] *Astrolabe* 2.8 deals with turning "the houres inequales in[to] houres equales," and 2.10 deals with how "to knowe the quantite of houres inequales by day." It seems that Chaucer was still having to cope with horological problems that the mechanical clock would eventually overcome. Although he makes many references to *Hours* in his poetry, the term (outside the *Astrolabe*) is used in a relatively imprecise sense. There are over twenty-five references to minutes in the *Astrolabe* and even two to seconds, but I have so far discovered no other references to minutes and seconds (either horological or astronomical) in the rest of Chaucer's canon.

Yet we know that, as a poet, Chaucer is intimately concerned with the passage of time. Despite his bourgeois background, which might have made more precise time measurement appealing, he is limited by his milieu. Time is measured by the return of the seasons or by the passage of day and night. Long poems—like *The Canterbury Tales* or the *Legend of Good Women*—open with an expression of pleasure at the return of spring (the *reverdi* typical of former poets). Of the shorter poems, *A Complaint to His Lady* takes its mood from the "longe nightes"; *The Complaint of Mars* is in the spirit of the equally widespread *Tagelied* or *Aubade,* in which the lovers regret the arrival of the day. The hours themselves are imprecise in early poetry. It would seem that even Chaucer cannot share with his audience a comprehension of minutes, let alone seconds, to indicate the fleeting nature of time.

As late as *Utopia* (1516), More was still using the term "equal hours," presumably in order to indicate a more precisely measured hour. He does so in the context of the Utopians who "divide the day and night into twenty-four equal hours and assign only six to work." But More also allocates the other hours very precisely, "not to waste in revelry or idleness."[10] The progress towards accurate time measurement and mechanical human activity is closely knit.

By the time Marlowe wrote his *Faustus* (1588?) clocks were sufficiently uniform in denoting a cycle of twelve hours twice daily for this concept to control the climax of the play. When "*The clocke strikes eleven*" (the eleventh hour), Faustus knows that he is almost damned after his twenty-four years of pleasure. He begs for "Perpetuall day, or let this houre be but a yeere,/A moneth, a weeke, a naturall day. . . ." But the movement of the clock is mechanical and inevitable: "The starres moove stil, time runs, the clocke will strike. . . ."[11] At eleven-thirty, Marlowe's stage instruction, "*The watch strikes*," suggests that there was generally no clear separation between the terms *watch* and *clock*. The first term derived from an alarm or "awaker," the second from the bell or "Glocke." Indeed, what we call a watch was earlier known as a pocket-clock.

The general scarcity of clock references in early English poetry[12] may well derive from the fact that horology seems to have been more advanced in Italy, France, and Germany. As Britten points out, "No English watch is known of a date before 1580 and only a short time before this is there any record of an English watchmaker,"[13] From then on, we may expect and do indeed find that horology makes an increasing impact on English literature.

Horological References in Shakespeare: The Poet Limited by His Age

That Shakespeare was, above all, the poet concerned with the passage of time is demonstrated by such sensitive studies as Quinones' *Renaissance Discovery of Time* and Turner's *Shakespeare and the Nature of Time*, but even Shakespeare, as we shall discover, was limited by his age. Like Chaucer, he had difficulty in denoting very small periods of time. However, unlike Chaucer, Shakespeare shares with his audience the concept of what a minute is like. He uses the term in this sense on more than sixty occasions, and, in *Macbeth*, even employs "minutely" as an adverb meaning "every minute."[14] Though a very few clocks with separate seconds dials did exist when Shakespeare was writing, this was not a concept that he could readily share with his audience. Insofar as I have been able to ascertain, the term *second*, used as a measurement of time, appears nowhere in the Shakespearian canon. Neither, though hours are frequently mentioned, does there any longer seem to be a need for specifying "equal" or "inequal" hours.[15]

Shakespeare's horological analogies are far from being limited to the mechanical clock; he also employs older artifacts, still in current use, and frequently adapts them to new concepts. The sandglass was now so often employed to measure equal hours that *hourglass* and *sandglass* were almost synonymous; as a result, Shakespeare could use the term *glass* to mean an hour. Thus when Ariel's "Past the midseason" is qualified by Prospero's "At least two glasses" (*Tempest* 1.2.239–40), the audience understands him to mean, "after 2 P.M." When Leontes talks about the condition under which his wife would not live "The running of one glass" (*Winter's Tale* 1.2.306), he means that she would not have an hour to live. In *I Henry VI*, the circumlocution, "ere the glass that now begins to run/Finish the process of his sandy hour" (4.2.35–36), means, "within the hour." So we know what Shakespeare intends to convey when, in the prologue to *Henry V* he talks metaphorically of "Turning the accomplishment of many years/Into an hourglass." In Shakespeare's sonnets, it is his "lovely boy" who paradoxically holds "Time's fickle glass, his sickle, hour..." (Sonnet 126). Sometimes the context may make one question whether even the term *clock* itself might not refer to a sandglass or a bell: "nimbler than the sands/That run i' the clock's behalf" (*Cymbeline* 3.2.74–75), or "To weep twixt clock and clock?" (*Cymbeline* 3.4.44).

The terms *dial* and *clock* were becoming relatively interchangeable, and can only be distinguished, if at all, by the context. When Lafeu's "dial goes not true" (*All's Well* 2.5.6), it is probably a watch; the dial that "points at five," in the *Comedy of Errors* (5.1.118), is probably a clock; and so is the "hourly dial" stopped by cogs in *Lucrece* (l. 327). But when Jaques meets a fool in the forest who "drew a dial" from his pocket and philosophized on the passage of the hours (*As You Like It* 2.7.20–33), this could just as well have been a pocket sundial as a watch. When time is to be told "by thy dial's shady stealth" (Sonnet 77), or when "yet doth beauty, like a dial hand,/Steal from his figure" (Sonnet 104), the context suggests that we are dealing with sundials rather than clocks. References like these to the sandglass, sundial, and clock suggest the type of problem facing historians of earlier horology. We have already referred to the difficulty in deciding when the mechanical clock first replaced the water clock and the manually motivated public bell some two hundred and fifty years before the time of Shakespeare.

Shakespeare rarely uses the term *watch* in its horological sense, but on two such occasions he clearly makes a pun between this and the more common optical use of the word (*Love's Labour's Lost* 3.1.191–95, and *Richard II* 5.5.52). The term *horologe* seems to be used only once. When Iago says that Cassio will "watch the horologe a double set," the reference is clearly to staying awake twice around a mechanical clock. While playfully allowing the Puritan Malvolio to daydream about marrying his mistress, Shakespeare conveys some of the flavor of importance attached to winding up one's watch (*Twelfth Night* 2.5.64–66). What Berowne has to say about clocks and women in *Love's Labour's Lost* demonstrates that (as in *Faustus*) the terms *watch* and *clock* were relatively interchangeable. It also demonstrates the general reputation of clocks. (Before the horological revolution, clocks were frequently corrected by sundials without any allowance for the sundial's error of up to sixteen minutes per day, related to the equation of time.) Berowne protests:

> What! . . . I seek a wife!
> A woman, that is like a German clock,
> Still a-repairing, ever out of frame,
> And never going aright, being a watch,
> But being watched that it may still go right! (3.1.191–95)

The unreliability of clocks gave rise to the phrase "the devil in the horologe," to which there are several early references. The term is used much like the "printer's devil" that has come down to modern times. After the late eighteenth century, Satanic qualities were sometimes connected with clockwork motivation and mechanical acts. I have been able to find no conclusive link between this and the earlier references to the devil in the horologe.[16] It is tempting to feel that there is a direct connection, too, between Shakespeare's woman "that is like a German clock," and Pope's contention that

> 'Tis with our *Judgments* as our *Watches*, none
> Go just *alike*, yet each believes his own.

The degrees of inaccuracy are relative; the very fact that one could compare watches, and anticipate that they might be similar, stresses the essential difference between the times of Shakespeare and of Pope. Pope would hardly have said of our judgements that they were "Still a-repairing, ever out of frame."

Sometimes a clock metaphor readily supplies concepts for which the ordinary vocabulary is barely adequate. We recognize immediately what Hermione means when she tells her husband that she loves him "not a jar [tick] o' the clock" less than another lady loves her lord (*Winter's Tale* 1.2.42–44); what Sebastian means when he says that Gonzalo is "winding up the watch of his wit./By and by it will strike" (*Tempest* 2.1.12–13); or what the king means, in *All's Well That Ends Well*, when he introduces the concepts of order and self-regulation that were later to be associated with the clock, "his honour,/Clock to itself, knew the true minute" (1.2.38–39).

Even the interrelationship between the mood of the man and the mood of the clock that we shall find to be essentially Wordsworthian and Dickensian is already foreshadowed in Shakespeare. In *The Winter's Tale*, lovers try to impose their feelings on the cold mechanism of horology, "Wishing clocks more swift?/Hours, minutes? Noon, midnight?" (1.2.289–90); in *King John*, old Time himself becomes "the clock-setter, that bald sexton Time" (3.1.324); and when the "*Clock strikes*" in *Twelfth Night*, Olivia says, "The clock upbraids me with the waste of time" (3.1.141). Iachimo, in *Cymbeline*, says of a past evil, "unhappy was the clock/That struck the hour!" (5.5.153–54).

We have noted some of the ways in which Shakespeare's references to horology and his use of horological analogies reflect the technological limitations of his age. But we have not yet considered Shakespeare in relation to some of the most important uses to which the horological analogy would be put. In this respect, Shakespeare's canon offers us the best possible gauge for the early use of horological metaphors in England. His career covers the first generation of English watchmaking, and he made his farewell speech in *The Tempest* barely a year after Galileo's use of the telescope provided the impetus to research that culminated in the horological revolution.

When Shakespeare retired, Descartes—whose mechanistic philosophy gave horological metaphors a new focus and importance—was fifteen years old. The basic metaphors which the mechanistic philosophy and those it influenced would employ were concerned with the element of order and regularity in a watch. Shakespeare was concerned as much as anyone with the problem of order,[17] but insofar as I have been able to ascertain he does not connect this with clocks.

The essential horological metaphors had all been used, albeit infrequently, before the time of Shakespeare. They were the analogies of the clock with animals, with men, and with the universe. Even the argument from design, using a clock analogy to prove the existence of God, had been made by Mornay in a work translated by Sir Philip Sidney and Arthur Golding and published in England in 1587. But this would take even longer than the other clock analogies to become established. One can only assume that the time was not yet appropriate.

I have been unable to find any clock analogies clearly made by Shakespeare that are related either to order in animals or to order in the universe. The case with human beings is somewhat different. Shakespeare does not relate people to clocks in the absolute sense that, for example, La Mettrie would later do in *L'Homme Machine.* Nor are people and clocks "yoked by violence together" as occurs in the Metaphysical poets. But Shakespeare goes further than Froissart or Visconti, who see in the clock merely another opportunity for a courtly love conceit. Without comparing men directly to clocks, he uses the fact that either as a whole or in part they have attributes which can be better explained by a comparison with clockwork.

In the first act of what is probably his first play, Shakespeare is proud to say of the English soldiers, who are attacking the French, "Their arms are set like clocks, still to strike on" (*I Henry VI* 1.2.42). The reference is almost certainly to clocks with single hour hands, and the image refers essentially to the mechanical nature of the clock. But there are none of the pejorative conotations that will be associated with clockwork automatons in the wake of the horological revolution.

When Shakespeare does use an extended horological conceit relating man to a clock, he is naturally more subtle than Froissart. At the end of *Richard II* at the moment of regicide which motivates the two tetralogies of history plays, the king resents being made Bolingbroke's puppet or "Jack-o'-the clock" (5.5.50–60). One hesitates to speak of coincidence in Shakespeare, but, at a similarly crucial point at the end of the other tetralogy, Richard III makes the fatal mistake of calling Buckingham a "Jack [of the clock]" (4.2.117). Richard II's reflection on time should be compared with the over-elaborate conceit on the same subject by the early Shakespeare, whose Henry VI carves "out dials quaintly, point by point,/Thereby to see the minutes how they run" (*III Henry VI* 2.5.24–40).

Shakespeare uses a horological conceit of a different sort when, in *As You Like It*, he considers the varying pace at which time's "foot" travels, even though "There's no clock in the forest." Rosalind amusingly explains to Orlando the differing speeds of time. It goes slowest for a maid between engagement and marriage; it ambles for a rich man and "a priest that lacks Latin"; it gallops for a thief whose time for hanging comes all too quickly; but it stops altogether for a lawyer who can "sleep between term and term" (3.2.317–51). Earlier in this play, Jaques had made the extended parallel, previously referred to, between the hours of the day and the ripening and rotting of man's life (2.7.20–30). Such references, together with the "eleventh hour" in *Faustus* indicate how clocks were able to provide an analogy with the progress in man's life. But the more normal poetic analogy for this was with the seasons.

In the speech which Bolingbroke concludes with the famous line, "Uneasy lies the head that wears a crown," he has been complaining that the god of sleep permits even the poorest people to be hushed into slumber "with buzzing night flies."

Why then, he asks, does slumber leave "the kingly couch/A watch case or a common 'larum bell?" (*II Henry IV* 3.1.16–17). No one seems to have been able to give a satisfactory explanation for this line,[18] but the influence of horology is clear enough.

The development of Bolingbroke's son Hal is from an apparent inattention to time and duty towards the sense of order that is necessary in a king. Hal, by the time he becomes Henry V, is a balanced king. He is not a man whose life, like the overhasty Hotspur, can "ride upon a dial's point" (*I Henry IV* 5.2.84), nor yet one who, like Falstaff, can sleep while the sheriff is seeking him. But, as early as *I HENRY IV*, Hal upbraids Falstaff (on one level the baser part of Hal that he must eventually disown) with inattention to time. What have you to do with the time of the day, Hal asks Falstaff, "Unless hours were cups of sack, and minutes capons, and clocks the tongues of bawds, and dials the signs of leaping houses [brothels], and the blessed sun himself a fair hot wench..." (1.2.7–11)? The suggestive use of parts of a clock was inevitable. Elsewhere, Shakespeare tells us that "the bawdy hand of the dial is now upon the prick of noon" (*Romeo and Juliet* 2.4.118–19). In much the same vein, Dryden, after Huygens had added a new dimension to clocks, describes how an "alluring girl" with her "lascivious eye" arouses the "lumpish pendulum."[19]

Though Shakespeare, as we have seen, uses the clock metaphor for many aspects of man, he does not yet develop the concept of man's body as a clockwork machine. Perhaps the essence of the change to come can be demonstrated by two metaphors, one from Shakespeare's *Othello* (3.3.355) and the other from Rochester's *Satyr against Mankind* (1. 29). During the horological revolution, Rochester's "reas'ning *Engine*" clearly refers to a man, but Shakespeare's "mortal engines" are quite different. They are, in fact, cannon.

The Use of Clock Analogies after Shakespeare

Before considering to what extent the poets after Shakespeare used clock analogies for explaining the universe, animals, and men, we shall look briefly at some of the more general uses for the metaphor. The very topicality of the clock partly accounts for the extent to which poets, philosophers, and theologians throughout the seventeenth and for much of the eighteenth century tend to use this analogy for explaining concepts that are essential to their particular interests.

Shakespeare's comparison between the vagaries of a German clock and a woman takes on a further dimension in Donne's

> Alas, we scarce live long enough to try
> Whether a true made clock run right, or lie.[20]

In a similar vein—and probably the predecessor to Pope's more famous statement—is Suckling's

> But as when an authentique Watch is showne,
> Each man windes up, and rectifies his owne,
> So in our verie Judgements. . . .[21]

The surprisingly mature young Pope tells us, at the beginning of his *Essay on Criticism* (1711), that our judgements are like our watches, and though none go quite alike each believes his own.[22] The fact that we are now concerned with relatively accurate watches is perhaps suggested by Dr. Johnson's contention towards the end of the century: "Dictionaries are like watches, the worst is better than none, and the best cannot be expected to go quite true."[23]

But judgements and dictionaries were not the only things that horology was expected to illustrate. In 1630, the master of Emmanuel College, Cambridge, felt that it might be suitably compared to religion, "in this curious clocke-worke of religion, every pin and wheele that is amisse distempers all: And as we are wont to lay aside cracked vessels, and distempered watches as unusefull, so doth God distempered and mixt religions."[24] In 1679 the "Rectour Hadham" uses the simile with pejorative connotations when he says of the Pharisees: "Their Religion was a kind of clock-work."[25]

Lord Chesterfield felt that a repeater watch might be compared to erudition, and advised his son: "Wear your learning, like your watch, in a private pocket; and do not . . . pull it out and strike it merely to show that you have one."[26] Dryden felt that clocks could explain the preoccupation with property (to be reflected, among others, by Locke's Whig values): "By these the springs of property were bent,/And wound so high, they cracked the government."

There were those who felt that the government itself could be explained in terms of clocks. Pope says, "Perhaps it may be with states as with clocks, which must have some dead weight hanging at them, to help and regulate the motion of the finer and more useful parts." Pope is usually doubtful about the complexities of government: "The nicest constitutions of government are often like the main pieces of clock-work, which depending on so many motions, are therefore more subject to be out of order."[27]

In the period after the horological revolution, Southey relates the analogy to the sense of automatism and predestination that came to be associated with clockwork: "The hand of the political horologe cannot go back, like the shadow upon Hezekiah's dial; . . . when the hour comes it must strike."[28] An essay in *Blackwoods* refers to "the whole machinery and watchwork of pauperism." Even when life comes to an end it frequently does so in mechanical terms related to a clock. Browning, in *Rabbi Ben Ezra*, declares that when "Time's wheel runs back or stops: Potter and clay endure."[29] But the constant use of clock analogies eventually tended to give them a certain staleness compared with those employed in the period leading up to and during the horological revolution. The subject was then so topical that it became the analogy most readily used for explaining abstract concepts in terms of a concrete mechanism understood by writer and reader alike.

Perhaps the most daring of such analogies is one used by John Donne in "Epithalamions, or Marriage Songs." He compares the undressing of the bride by her ladies on the wedding night, "as though/They were to take a clock in peeces."[30] Little wonder that Dr. Johnson described the Metaphysical poets as those by whom "the most heterogeneous ideas are yoked by violence together."

In the early years of the eighteenth century, the Scriblerians (who included Pope and Swift) could hardly avoid being fascinated by clocks and employing them for the purpose of analogy. Pope, for example, uses a variation of the following couplet in three of his works:

> So Clocks to Lead their nimble Motions owe,
> The Springs above urg'd by the Weight below.[31]

The chiming of numerous clocks in aristocratic homes is typical of Pope's age. In Timon's villa, "the chiming Clocks to dinner call" (3.2.152); in Belinda's home, in the first version of *The Rape of the Lock*, "striking Watches the tenth Hour resound" (2.127). When he revised *The Rape*, however, Pope introduced the even more topical repeater, which his affected Belinda presumably presses in broad daylight, in order to learn the time: "the press'd Watch return'd a silver Sound" (2.146).

But Pope was interested in more than the social aspects of horology. He also recognized the influence of clocks on industry. We have earlier discussed Tompion's use of the division of labor in watchmaking. Both Pope and Mandeville applied this topical aspect of technology to similes that reinforce essential arguments in their work. Mandeville, in his *Fable of the Bees*, argues that the division of labor will grow directly out of the fact that peace in society permits men to specialize. In the "Sixth Dialogue," Horatio agrees with Cleomenes on this point: "The truth of what you say is in nothing so conspicuous, as it is in Watch-making, which is come to a higher degree of Perfection, than it would have been arrived at yet, if the whole had always remain'd the Employment of one Person; and I am persuaded, that even the Plenty we have of Clocks and Watches, as well as the Exactness and Beauty they may be made of, are chiefly owing to the Division that has been made of that Art into many Branches."[32]

Pope combines the concepts of the division of labor and the "manufacturer" who "puts all together" in his "Project for the Advancement of the Bathos." He begins by telling us that "the vast improvement of modern manufactures ariseth from their being divided into several branches, and parcelled out to several trades: For instance, in clock-making one artist makes the balance, another the main spring, another the crown-wheels, a fourth the case, and the principal workman puts all together: to this economy we owe the perfection of our modern watches, and doubtless we also might that of our modern poetry and rhetoric, were the several parts branched out in the like manner."[33] Pope then amusingly proceeds to indicate how experts in hyperbole, periphrasis, proverbs, and a number of other specialities the misuse of which have received individual attention in *Peri Bathous* could contribute to a common stock from which a master "poet" or "orator" might assemble his work.

One could adduce many other examples of the ways in which poets in the eighteenth century reflected the influence of horology not only on society, including themselves, but also on industry.

The Use of the Clock Analogy to Explain the Nature of the Universe, of Animals, and of Man

We have noted how poets used the clock analogy to explain concepts that were relevant to their respective interests. There was also an interest which all poets had in common: the nature of the universe, of animals, and of man. Oresme had compared the universe to a clock before the end of the fourteenth century; but it was not until three centuries later, and more particularly until the time of the horological revolution, that this analogy began to be regularly employed. The clock analogy for the universe seems to have been valued most for the sense of order that it implied. If this is so, clocks would have to be sufficiently accurate and sufficiently popular for the analogy to be generally appreciated. Glanvill's translation of Fontenelle's *Plurality of Worlds* (1686, translated 1688) brought to Englishmen a popular introduction to Cartesian ideas: "I perceive, *said the Countess*, Philosphy is now become very Mechanical. So mechanical, *said I*, that I fear we shall quickly be ashamed of it; they will have the World to be in great, what a Watch is in little." But the Countess defends this mechanical universe, "I value it the more since I know it resembles a Watch, and the whole order of Nature the more plain and easie it is, to me it appears the more admirable."[34] The concept of order and regularity inherent in the mechanism of a watch is what gave the clock analogy its greatest value during the neoclassical period. Later this very rigidity proved an important factor in the association of diabolism with clockwork.

Copernicus' heliocentric cosmos resulted in astronomers' demanding clocks of far greater accuracy. But the work of Copernicus also produced a potential for questioning the order in the universe, the assumptions of religion, and even the existence of God himself. Ironically the new clocks were to provide the analogy through which the new questioning of order would be effectively combatted during a period of some two hundred years.

In the period with which we are concerned, the universe was compared to a clock, and, as a corollary, God became a sort of celestial watchmaker.[35] The concepts of the universe as a clock and its God as a watchmaker belong more properly to the history of philosophy and theology respectively, which were discussed in chapters 4 and 5.[36] Our purpose now is to demonstrate briefly how wide a spectrum of other writers made reference to the clock universe during the horological revolution. In his *History of Most Manual Arts* (1661), the anonymous author compares the watch to the universe, "the *wheel-work* of this great Machin." In Power's *Experimental Philosophy* (1664), he argues that "he that made this great Automaton of the world, will not destroy it" until—just as would "a common Watchmaker . . . that has made a curious Watch"—the universe is allowed to work through all its revolutions at least once. Happily, Power estimates that since this process would take Twenty thousand years, the universe still had fifteen thousand years left.[37] Not only natural philosophers, but also mystics employed the clock analogy for the universe. Taylor, when he "unfolded" *Jacob Behmen's Theosophick Philosophy* (1691), refers to "the Soul of the World, like a Horologe" and "this Soul the Horologe of Nature."[38] Vaughan, who was both a clergyman and a poet, writes that "Heav'n/Is a plain watch, . . ." and mentions elsewhere "The hours to which high Heaven doth chime."[39]

In the middle of the eighteenth century, Fielding—needing an example for demonstrating that "the greatest events are produced by a nice train of little circumstances"—understandably turns to the watch. Swift had used a comparable image in the first two paragraphs of *Tale of a Tub* (section 9): "What secret Wheel, what hidden Spring could put into Motion so wonderful an Engine?" Fielding, speaking of the macrocosm rather than the microcosm, tells us that "the world may indeed be considered as a vast machine, in which the great wheels are originally set in motion by those which are very minute, and almost imperceptible to any but the strongest eyes."[40]

The above are just a few examples of the very many ways in which writers of all sorts, including poets, compared the universe to a clock. But poets and philosophers seem to part company when it comes to using the clock analogy for animals and for men. Poets appear prepared to use the metaphor for man, whereas philosophers only do so with the greatest circumspection. On the other hand, while philosophers frequently compare beasts to clocks, in the spirit of the mechanistic philosophy of Descartes, poets seem unable to treat that subject seriously.

In Richard Leigh's *Poems* (1675), he writes of insects:

> Like *Living Watches*, each of these conceals
> A thousand *Springs of Life*, and *moving wheels*,

but I have been hard put to find another example in poetry. Even here the stress on the word *living* may be intended to question a completely Cartesian view. There are many examples of poets mocking at such concepts, and we shall consider these in the next chapter when dealing with their reaction to Cartesian mechanism.

Perhaps one explanation for the difference between the viewpoints of poets and philosophers is that the latter felt a greater obligation to carry through their analogies in terms of a total system. For physiological and later psychological studies it was convenient to explain the operation of animals in terms of clockwork. In this respect, it was safer to assume that they had no souls. But poets, like some philosophers, could not help noting that animals were more like men than Cartesians dared allow. Also, there was a distinctly atheistic potential in the Cartesian dualism that sought to suggest that man's mechanical body operated in conjunction with his eternal soul. At least until the time of Sterne's *Tristram Shandy*, poets continued to laugh at the pineal gland through which Descartes attempted to organize this latter-day miracle.

But poets—perhaps precisely because they did not have to carry their metaphors through to a logical conclusion—did use the topical clock analogy for explaining men. It is possible to argue that the mechanical analogy of the watch was grafted onto the older concepts from a premechanical age. Descartes' dualism, for example, is comparable to the relationship between the body and the soul with which theologians and poets had been concerned for a long time. Also the frequently asserted relationship between macrocosm and microcosm permitted an analogy applicable to a world or a universe to be transferred to "the shape of man, which may be called a little world."[41]

John Donne employs such concepts to structure his thought. In his *Obsequies to the Lord Harrington*, he employs an extended horological simile that compares small people to watches whose faults "only on the wearer fall," and great people to great clocks whose errors affect many. But he goes beyond this in feeling that Harrington's soul has a clock so true and so closely regulated by God that it can both control the sun, and act as a great "sun dyall" by which all the other people and clocks might, in their turn, be regulated. We may note that the soul in Donne's "pocket-clock" is the spring. There are no Cartesian pineal glands separating it from the rest of the wheel work.[42]

The Metaphysical poets of the seventeenth century were fascinated with the new developments in science and technology. Though some—like Donne, Herbert, and Vaughan—were churchmen themselves, they did not avoid using the most daring analogies. It is perhaps because they reflect strongly the dichotomy between technology and religion that the Metaphysical poets have been so popular in our own age.

Descartes and his follower Clauberg attempted later to explain the difference between a living man or beast and a dead body by comparing them respectively to a clock in running order, and a clock that has stopped. Donne and Herbert seem to have anticipated this by a similar analogy. In a *Funerall Elegie*, Donne compares the dead body to "a sundred clocke" temporarily "peecemeale laid." (It is typical of Donne that the only other person he compares to taking a "clock in peeces" is a bride being undressed on her wedding night. She, too, is about to "die" in terms of a popular Elizabethan pun that Donne frequently uses.)[43] When Herbert is addressing his Maker, also on the subject of death, he too anticipates the day when God will "give new wheels to our disorder'd clocks."[44]

Vaughan seems to play with the idea that although "Heav'n/Is a plain watch," time only affects those on earth. This permits him to produce a paradox typical of the Metaphysicals: "The last gasp of time/Is thy first breath, and man's *eternall Prime.*"[45] The analogy of God as watchmaker is, of course, implicit in the examples quoted from all three Metaphysical poets. That was something they could share with philosophers and theologians alike.

Herrick, in "The Watch," says categorically, "Man is a Watch"; and, in *Aeropagitica*, Milton feels that without reason and the ability to choose Adam would have been a mere puppet.[46] Earlier, when writing on the death of Hobson—the famous university carrier at Cambridge who has given us the term *Hobson's choice*—the younger Milton likened him to a clock that has now stopped.[47] When, in this period, Milton wrote Lycidas—as an elegy for his Cambridge friend, Edward King—he included a couplet whose meaning has long tantalized scholars. St. Peter says:

> "But that two-handed engine at the door,
> Stands ready to smite once, and smite no more."

Suggestions for the "two-handed engine" include St. Peter's keys, a large two-handed sword, and the two Houses of Parliament. Surely the two-handed engine as a man or as Time or Death personified is a possible connotation which adds further value to that haunting image.[48] Milton's own Hobson is "like an Engin," and certainly by the time that Rochester talks of the "reas'ning Engine!" in 1675, we know that he is referring to a man.

One aspect of the reaction of the poets to the concept of man as clockwork is not overtly shared by either the philosophers or the theologians. This was the reaction of the poets to clock-dominated bourgeois values. Walter, the retired "*Turkey* merchant" in Sterne's *Tristram Shandy*, is mercilessly undercut for his clock-oriented actions and his mechanical writings. But the true bourgeois was proud of his regularity. It is for this, above all, that Richardson, the printer, lets Pamela praise Mr. B. After listing the remarkably precise hours that he keeps, Pamela says approvingly of her husband, "He is a regular piece of clockwork! . . . why . . . should we not be so? For man is as frail a piece of machinery as any clock-work whatever: and, by irregularity, is as subject to be disordered."[49] Kings, too, became regulated by the watch during the horological revolution. Saint Simon said of Louis XIV: "Avec un almanach et une montre, on pouvoit dire, à trois cents lieues de là, ce qu'il faisoit."

If man is a watch then the wheels of life are watch wheels. Dryden, in *Oedipus* 4.1, employed the simile: "Till, like a clock worn out with eating time,/The wheels of weary life at last stood still"[50] Armstrong, in *The Art of Preserving Health* (1744), talked of "the hesitating wheels of life." Like so many metaphors that either derive from the clock, or have been adapted for use with it, the "wheels of life" have now passed into our language. In the nineteenth century, Bulwer-Lytton took this analogy into man's social and physical existence. For the poor, any "suspense of health" means that "all the humble clock-work is undone."[51]

Byron, using the synecdoche that Pope deplored, facetiously suggested that man was no more than a watch part, a "pendulum betwixt a smile and a tear." But then Byron was, in some measure, a latter-day Augustan. At the beginning of the *Vision of Judgement*, he poked gentle fun at the Newtonian watch-universe which, unlike that of Leibniz, needed regulating occasionally.[52]

It is an irony, already noted, that whereas poets seemed to be able to employ for men the same mechanical analogy that they used for the universe, philosophers did so at their peril. Hobbes, Spinoza, Hartley, and La Mettrie all risked being branded as atheists for even the most circumspect suggestions that men operated according to mechanical laws. Yet, in the nineteenth century, Hazlitt and Holmes appear to have been able to transfer the mechanical analogies of Newton and Leibniz for the universe into comparable analogies for the mind. Hazlitt reflects the

Newtonian universe when he says, "The mind of man is like a clock that is always running down, and requires to be as constantly wound up."[53] Oliver Wendell Holmes, on the other hand, reflects the macrocosm of Leibniz: "Our brains are seventy-year old clocks. The Angel of Life winds them up once for all, then closes the case, and gives the key into the hand of the Angel of Resurrection."[54]

During the horological revolution, the metaphor of man as a clock sometimes slipped almost unnoticed into the language. In Congreve's *Double Dealer* 1.1, Lady Touchwood, in three successive speeches, plays with the idea that Maskwell, the villain, is trying to "wind" and "unwind" her "like a Larum." Pomfret's popular poem *The Choice*—which exemplified the Augustan yearning after an urbane country life—commends the "Sweetness in a Female Mind" that "Winds up the Springs of Life."[55]

Although the proto-Romantics and Romantics generally reacted unfavorably to the quality of order inherent in the clock, the analogy with man sometimes continues. For Cowper, "An Idler is a watch that wants both hands," and Landor calls man a "breathing dial."[56]

After the eighteenth century, philosophers and scientists tended to turn away from clock metaphors. Poets and theologians continued to employ the analogy,[57] but in different ways. For poets, with whom we are here primarily concerned, the watch frequently takes on certain human qualities. It can suggest (as foreshadowed in Shakespeare) something derived from the senses of the poet, reflecting "what they half create,/And what perceive." In Wordsworth's *Guilt and Sorrow*, the clock (reflecting the mood of the poet) tolls "dismally," just as it does in *The Idiot Boy*. It is a "solitary clock" in "*The Solitary*" of *Excursion* (book 2); a "loquacious clock" in Wordsworth's Cambridge days of *Prelude* (book 3); a "monitory clock" in "Blest is this Isle"; and a slow, deep clock in *An Evening Walk*.

In Dickens the interest in clocks is more pervasive. As with Wordsworth, Dickens' clocks can take on human qualities, but his interest is also much broader. There are three main aspects of the Dickensian obsession with clocks. He is sentimental about them; he deplores the Satanic quality of mechanical actions and a mechanical heart; and his lively imagination frequently envisages people like clocks or clocks like people. It is only with the latter, the clock analogies, that we are concerned at this point.

There are times when Dickens' people seem to be like clocks. In *Dombey and Son*, Dickens says of Perch, the messenger in Mr. Dombey's office, that his "place was on a little bracket, like a timepiece." As Mrs. Dombey lies dying, "the large watch that defines Mr. Dombey races with the Doctor's watch." When writing to his biographer, John Forster, Dickens describes a prefect of police on horseback, "turning his head incessantly from side to side like a Dutch clock." In *Hard Times*, the simile of a Dutch clock is used to illustrate the sound of Tom Gradgrind's snoring. Elsewhere Dickens says, "I can warrant myself in all things as punctual as the clock at the Horse Guards."[58] In *The Chimes*, Toby Veck's "Fancy" seems to be personified in the voices of the bells that dominate his life.

At other times, Dickens' clocks reflect, like Wordsworth's, the nature of people. They can become as pallid as the humans with whom they associate: in a baker's shop, in *The Uncommercial Traveller*, there is "a hard, pale clock"; Uriah Heep possesses "a pale, inexpressive-faced watch"; and, in a pawnbroker's shop, Edwin Drood sees "dim and pale old watches, apparently in a slow perspiration." (But in *Household Words* chronometers being tested at Greenwich seem "like so many watch-pies in a baker's dish.") On the Dutch clock in *The Cricket on the Hearth* stands a "terrified Haymaker . . . jerking away right and left with a scythe in front of a Moorish Palace." A secondhand furniture shop, in *Dombey and Son*, contains "motionless clocks that never stirred a finger and seemed as incapable of being successfully wound up as the pecuniary affairs of their former owners."

The clock analogy in Dickens appears very different from that used by Froissart and Visconti in the service of courtly love. Yet by employing the many variations of the analogy, particularly during the horological revolution, poets used clocks for the main purpose of poetry. Through them, the poet gave to man a heightened awareness of himself and his universe.

The Poets' Reaction to Cartesian Clockwork

The Ridiculing of Scientists for Undue Attention to the Clock

The main purpose of this chapter is to consider the poets' concerted attack, during the horological revolution, on the concept of Cartesian mechanism in animals. Since this must in some measure be seen as part of the larger satiric attack against scientists in the wake of the founding of the Royal Society, we should first look at the way in which scientists (as well as moderns and philosophers with whom they were associated by poets) were being mocked for undue attention to the clock. We shall, however, also demonstrate that not until Sterne's *Tristram Shandy*, in 1760, do we find a concerted attack on the clockwork nature of man.

In stage comedies, satire against scientists can be traced from Shadwell's *Virtuoso* (1676) to the middle of the eighteenth century, when science had achieved an honorable position among intellectuals.[1] But the undercutting was by no means restricted to stage comedies. In a curious pamphlet from the *Miscellanies* (1732), that has been attributed to Swift, the author mocks both the prophecies and the pedantic precision of the scientist: "But on Wednesday morning (I believe to the exact calculation of Mr. Whiston) the comet appeared; for, at three minutes after five by my own watch, I saw it. He indeed foretold, that it would be seen at five minutes after five; but, as the best watches may be a minute or two too slow, I am apt to think his calculation just to a minute."[2] Swift makes comparable use of mock horological precision elsewhere: "Looking upon my Watch, I found it to be above five Minutes after Seven: By which it is clear, that Mr. *Bickerstaff* was mistaken almost four Hours in his Calculation. In other Circumstances he was exact enough."[3]

The extent to which scientists could be ridiculed for being ruled by the clock had its parallel in the bourgeois habits that were becoming prevalent. Wycherley's *Country Wife* (1675) is remarkably topical in this respect. Sir Jasper Fidget—the city knight who typically doubles as cuckold—is obliged to leave his wife with the rake, Horner: "Nor can I stay longer; 'tis—let me see, a quarter and a half quarter of a minute past eleven; the Council will be sate, I must away: business must be preferr'd always before Love and Ceremony with the wise Mr. *Horner*."[4] The bourgeois cuckold trying to drag a flighty wife away from the ball by pointing at his watch in Hogarth's *Analysis of Beauty* (plate 2) tells a parallel story of the new slavery to time. Pope's Sir Balaam, the city knight, is "Constant at Church, and Change," while Cowper, in *Expostulation*, talks of "business, constant as the wheels of time."

At the end of the horological revolution, Sterne, as much concerned with mechanism as Swift, adapted the new accuracy of stopwatches to what was effectively an old joke:

> —And how did *Garrick* speak the soliloquy last night?—Oh against all rule, my Lord,—most ungrammatically!... he suspended his voice in the epilogue a dozen times three seconds and three fifths by a stop-watch, my Lord, each time.—Admirable grammarian!... Was the eye silent? Did you narrowly look?—I looked only at the stopwatch, my Lord.—Excellent observer![5]

It is today generally accepted that bourgeois and pedantic qualities are closely related to an undue concern with time. In Beckett's *Waiting for Godot*, the watch—"A genuine half-hunter, gentlemen, with deadbeat escapement"—that Pozzo carries around with him characterizes its owner.

During the horological revolution, there was very little reaction from the poets against the concept of the clockwork universe, and certainly not against its corollary the God who was a divine watchmaker. Sometimes, antagonism towards philosophers in general undercut their abstruse arguments about the universe. Swift frequently satirized the "Philosopher, who, while His Thoughts and Eyes were fixed upon the Constellations, found himself seduced by his *lower Parts* into a *Ditch*."[6] Pope, towards the end of *The Dunciad*, attacks those "such as Hobbs, Spinoza, Des Cartes, and some better Reasoners" who "take the high Priori Road,/And reason downward, till we doubt of God," thrusting instead "some Mechanic Cause into his place."[7] The Hudibrastic verses of Prior's long poem *Alma; or the Progress of the Mind* fairly interpret the poets' suspicion of philosophers, the "system-makers" who "fight as Leibnitz did with Clarke."[8]

However, the stature of philosophers like Leibniz and Newton (on whose behalf Clarke was fighting), together with the needs of theologians, would seem to have restrained the satire of the poets from making a direct attack on the conception of the clockwork universe. In the nineteenth century, by comparison, we find Clough directly questioning the "Mécanique céleste." (Laplace's *Mécanique céleste* [1799–1825] was a monumental work that summarized a century of research on gravitation.)[9]

Swift and Sterne: Changing Attitudes towards Attacking the Concept of Man as a Clock

Although there are some important exceptions, poets were not consistent in attacking the concept of man as a clock during the horological revolution. Perhaps this is because it was difficult to attack philosophers who were themselves wary of the dangers inherent in taking Cartesian mechanism to the extreme implied by La Mettrie's *L'Homme Machine* (1748). Berkeley's anonymous *Guardian* essays (nos. 35 and 39) on "The Pineal Gland" demonstrate the suspicion of philosophers regarding the gland through with Descartes rationalized contact between the mind and a mechanical clockwork body. The essays are written in a lighter vein than is Berkeley's custom; they deal with a box of snuff given to the author by an uncle who reputedly wrote *The Voyage to the World of Descartes.* The uncle was a virtuoso who invented a form of snuff that could separate body from soul, because Descartes, "having considered the body as a machine or piece of clockwork... began to think a way may be found out for separating the soul for some time from the body, without any injury to the latter."[10] It was the Cartesian dualism of body and soul that troubled most such theologically oriented philosophers as Berkeley.

Swift's *Discourse concerning the Mechanical Operation of the Spirit* (with which sections 8 and 9 of *Tale of a Tub* have been frequently associated) is better understood when the attack on mechanistic philosophy is taken into account. Though Swift frequently assails Descartes and Hobbes, his attack does not generally seem to be directed against the clockwork nature of man. However, kings, in the person of Henry of Navarre, are undercut through a watch analogy in *Tale of a Tub* (section 9), and the bourgeois clockwork nature of Gulliver seems to be satirized through the indirection of a report to the Lilliputian king. Of Gulliver's great silver watch it is said, "We conjecture it is either some unknown Animal, or the God that he worships: But we are more inclined to the latter Opinion, because he assured us... that he seldom did any Thing without consulting it. He called it his Oracle, and said it pointed out the Time for every Action of his Life."[11] Though Swift did not press this point, he seems to have been aware of some of the criticisms that would later be more consistently made against clockwork and clockwork-type men.

Sterne's *Tristram*, for good reason, acknowledges its indebtedness to the *Tale of a Tub.* Like Swift's early trilogy (*Tale*, *Battle*, and *Mechanical Operation*), it is much concerned with an attack on mechanistic philosophy. Nevertheless, there are some important differences. Descartes, his dualism, and even the pineal gland are attacked. But it is the mechanical process of association that controls the "digressive and progressive movements" of the book. *Tristram* itself became a mechanical operation related to the author-persona, in which "one wheel within another... the whole machine, in general, has been kept a-going."

Walter and indeed the whole Shandy family are controlled by the clock. There are also several compelling hints that the father's sterility is not restricted to his dry methodical hypotheses. An added irony is that when we learn about the clock, we learn also that Walter was a retired Turkey merchant. In the second half of the eighteenth century, Turkey and China were two of England's most important export markets for unusual clocks and watches.[12] This particular twist of Sterne's satiric knife would hardly have escaped contemporary readers.

We know that Sterne reacted favorably to the first of the many pamphlet attacks on *Tristram*,[13] quite possibly because he had written it himself. The pamphlet was entitled *The Clockmakers Outcry against the Author of the Life and Opinions of Tristram Shandy* (1760). The "clock maker"-author expresses a "just indignation for what we and our brethren the clockmakers suffer through the heretical and damnable Opinions of *Tristram Shandy*." The author prefers the "decently entertaining manner" in which Prior had dealt with Descartes' pineal gland in *Alma* to the "libidinous images" liable "to knock up all order" and bring "the works of our fraternity into disgrace." He refers in particular to the connotations of Tristram's mother asking Walter whether he has "*not forgot to wind up the clock*?" The author notes pertinently that Walter (Widow Wadman's late husband suffered from the same disease) had "the Sciatica" in the months when Tristram was presumably conceived, and therefore could "not so much as wind up the clock. The author ought to have told the reader who wound it up in his stead."[14] As many ardent Shandeans are aware, Toby, Yorick, Obadiah, and Slop are all possible contenders, but Sterne continues to keep us guessing.

Apart from valid insights into *Tristram* which are not the subject of our present consideration, the pamphlet is also involved in some gentle mockery of the bourgeois and pedantic nature of the clockmakers' club (rather like a mechanics' equivalent of the Spectator's Club). The *Clockmakers Outcry* ends on a particularly amusing note when one of the members assures his fellows that "this *Tristram*, as I have learned by letters from the country, is like to ruin our trade." Understandably, "At this they all looked grave." The clockmakers feel that the connotations attached to "winding up the clock" will undermine church and state, and bring much the same chaos that Pope had forecast in *The Dunciad*. This humorous passage effectively undercuts both clockmakers and the clockwork nature of men. In 1760, it was possible to do this much more successfully than at the height of the horological revolution:

> The directions I had for making several clocks for the country are countermanded; because no modest lady now dares to mention a word about *winding-up a clock*, without exposing herself to the sly leers and jokes of the family, to her frequent confusion. Nay, the common expression of street-walkers is, "Sir, will you have your clock wound-up?"[15]

Sterne's humor—if Sterne is the author of this pamphlet that ran to four editions in 1760—would appear to be founded on the situation in which a relatively modest trade has gone through an affluent period, and some of its members are showing a certain amount of affectation. It is very difficult to recapture this type of social history, but two extracts from literature may help. In the preface to the first edition of Boileau's *Lutrin* (1674), the following statement is revealing: "C'est un Burlesque nouveau, dont je me suis avisé en nostre Langue. Car au lieu que dans l'autre Burlesque Didon et Enée parloient comme des Harangeres et des Crocheteurs; dans celui-ci une Horlogere et un Horologer parlent comme Didon et Enée."[16] Here the watchmaker seems to be used as the typical example of a relatively humble trade. But by the end of the horological revolution the situation was different. Johnson's popular history of Betty Broom in the *Idler* (1758, nos. 26 and 29) was reprinted in at least six other journals. Betty's first employer was a watchmaker whose affections and afflictions both derived from being able to earn money too easily.[17]

In *Tristram Shandy*, Walter—the Cartesian (3.18) and retired Turkey merchant—represents the evils deriving from a clockwork mentality. It is he who installed the clock; he who insists through a subtle allusion to the penis that Descartes' "*pineal* gland" cannot be in the brain (2.19, and "Slawkenbergius's Tale");[18] and he who feels that the method for "counterbalancing evil" is not Toby's religion but "a great and elastic power within us . . . like a secret spring in a well-ordered machine" (4.8). There is a masterstroke of irony in Sterne's giving Walter the line: "I wish there was not a clock in the kingdom" (3.18).[19]

We have previously taken note of the critic who, in *Tristram*, judged Garrick by measuring the duration of his pauses on a stopwatch. In a remarkably parallel passage, Tristram defends himself against those who may criticize his idiosyncrasies with chronology. (The novel is humorously "structured" on the erratic use of Locke's mechanical association of ideas.)[20] In *Tristram* 2.8, Sterne seems to be saying—and 1760 is an appropriate time to say it—that the clockwork control of literature is no longer acceptable:

> If the hypercritic will go upon this; and is resolved after all to take a pendulum, and measure the true distance betwixt the ringing of the bell, and the rap at the door;—and, after finding it to be no more than two minutes, thirteen seconds, and three fifths,—should taken upon him to insult over me for such a breach in the unity, or rather probability, of time;—I would remind him, that the idea of duration, and of its simple modes, is got merely from the train and succession of our ideas,—and is the true scholastic pendulum,—and by which, as a scholar, I will be tried in this matter,—abjuring and detesting the jurisdiction of all other pendulums whatever.

Sterne undermines the clockwork nature of man, and the clockwork chronology through which his actions are normally represented in literature. He also undermines the mechanical nature of the literary forms to which the Augustan period subscribed: "I should beg Mr. Horace's pardon;—for in writing what I have set about, I shall confine myself neither to his rules, nor to any man's rules that ever lived." Sterne's sentiment and his disregard for the unities belong to the latter part of the century. His hero's life is molded by the clockwork head of Walter and the feeling heart of Toby.

Though Sterne's work is influenced by the nature of his age, it would be a mistake to bracket him with Johnson or Richardson. He belongs rather to the tradition of Rabelais, Cervantes, and Swift, to whom he pays tribute. Sterne had adapted his works to the requirements of a new age by ridiculing what already existed rather than by producing a new organic unity of his own. Future attacks on clockwork men will tend to be involved and subjective rather than satiric and objective—the total work will be more in harmony with the aesthetic of a new age. But this is to anticipate what we shall be saying about the clockwork devil.

The Horological Revolution: Relatively Few Attacks by Poets on the Clockwork Nature of Man

During the horological revolution, there were relatively few attacks made on the clockwork nature of man. Though these tended to be in the satiric mode, none of them could in any way be compared with the concerted assault that Sterne was able to mount after 1760. In the following passage from *Hudibras* 3.1, Butler appears to have associated a pun on the name of Descartes (Des Cartes) with criminals who hang like watch pendulums:

> Without capacity of Bail
> But of a *Carts* or *Horses Tail*:
> And did not doubt to bring the Wretches,
> To serve for *Pendulums to Watches*:
> Which modern Vertuoso's say,
> Incline to Hanging every way.

In 1678, when the third part of *Hudibras* was written, this was particularly topical, since pendulum watches were a temporary phenomenon rendered obsolete by the discovery of the balance spring (c.1674).[21] Pendulum watches suffered from problems similar to those of pendulum clocks at sea, and therefore inclined to "Hanging every way."

Pope, in *Martinus Scriblerus*, attacks more directly the concept of a mechanical mind. He allows the virtuosos (amateur scientists who were frequently the butt of satire) to describe the operation of the brain in mechanical terms, and then explains how they have arranged for a clockwork-motivated "artificial man" to be constructed. The satiric stance tends to stress the stupidity rather than the evil of the virtuosos; certainly no evil (as occurs with the later Romantic "Frankensteins") appears, as yet, to adhere to the clockwork man himself: "We are persuaded that this our artificial man will not only walk, and speak, and perform most of the outward actions of the animal life, but (being wound up once a week) will perhaps reason as well as most of your country parsons."[22]

Explaining man through a clockwork metaphor becomes considerably more difficult when questions of growth and decay are involved. In his *Essay concerning Human Understanding*, Locke uses an extended clock metaphor in his attempt to explain the difference between species ("Nobody will doubt that the wheels or springs [if I may so say] within are different in a *rational man*..." [3.6]). He also uses an extended watch metaphor in his attempt to explain identity. The analogy is between "a colt grown up into a horse, sometimes fat, sometimes lean," which "is all the while the same horse," and a watch, "whose organized parts were repaired, increased, or diminished by a constant addition or separation of insensible parts, with one common life..." (2.27). Clearly the clock provides only an awkward analogy for organic growth.

Prior may have Locke's analogy in mind when he discusses at length "the goings of this clock-work man" in *Alma*: the differences between men—represented by auxiliary movements in the watch—are not essential, but without food, the essential "horal orbit" ceases and the watch stops.[23] In the nineteenth century, Browning, faced with the problem of growth, could simply deny the watch analogy. For him, man was no longer "made a wheel work to wind up."[24] But early in the twentieth century, Hare's amusing limerick puts the clockwork metaphor into a modern idiom. Even man's mechanical movements are now mechanically confined:

> There once was a man who said, "Damn!
> It is borne upon me that I am
> An engine that moves
> In predestinate grooves,
> I'm not even a bus I'm a tram."

The Horological Revolution: The Virtual Unanimity of Poets in Attacking the Cartesian Concept of Clockwork Animals

We have found that although philosophers were very circumspect about treating men as clockwork, poets both used and abused that concept. Cartesian and other philosphers, on the other hand, made considerable use of the clockwork analogy in explaining animals. But for this aspect of the mechanistic philosophy poets had nothing but constant and continuing abuse.

Leonora Cohen Rosenfield—who traces the development in France "From Beast Machine to Man Machine"—names only Louis Racine, the undistinguished literary son of the French poet, as a consistent supporter of animal mechanism. There, too, the literary opposition was vociferous, but by 1737 even Macy, a Cartesian, had acknowledged the discrediting of the automatists' view of animals. In 1747 the Cartesian poet Louis Racine reaffirmed the same status of defeat.[25] Fontaine, though critical, confirms how considerable an impact the ideas of Descartes made at Port Royal: "Il n'y avoit guère de solitaire qui ne parlât d'*automate*." Dogs, because they were nothing but watches, could be beaten or opened up "tout en vie." Elsewhere he says, "le systême de Descartes sur les bêtes, soutenoit que ce n'étoient que des horloges, & que quand elles crioient ce n'étoit qu'une roue d'horloge qui faisoit du bruit."[26] Samber's translation of La Motte's *Fables* (1721) makes a comparable attack on animal mechanism: "Nothing but bold intrepid *Cartesianism* could make it a Matter of Dispute; but it is, perhaps, a depraved Way of Reasoning that could dare to make them meer Machines, or pieces of Clock Work."[27] It is interesting to note that the adjective "mere" and the phrase about clockwork have been added by the English translator. The French version of 1719 reads simply "d'en avoir osé faire des machines." The attacks of Fontaine and de la Motte employ polemic rather than ridicule.

The tone of Fontenelle and Bougeant is lighter but equally effective. Fontenelle's lettre 9, "A Monsieur C," was introduced to England through an apparently unacknowledged translation in Thomas Brown's highly popular works: "You pretend, that beasts are no less machines than watches. Now, I dare engage, that if you put a certain machine, call'd a dog, and another machine, call'd a bitch, together in the same room, there will result a third little machine from their corresponding together; whereas you may keep two watches together as long as you live, nay, 'till doomsday, if you please, and they will never produce a third watch between them." The writer says that their mutual friend Madam B—— regrets that Monsieur C—— has lost his reason, "I dare swear she would strangle *Des Cartes*, in one of her garters, if she had him in the room." The letter ends in the same vein: "As for me, I assure you, I am a piece of clock-work new wound up, to go into your service...." At least nine French editions of this letter appeared between 1683 and 1742.[28]

Father Bougeant's attack on Cartesian mechanism also appeared in English translation, "I defy all the *Cartesians* in the World to persuade you that your Bitch is a meer Machine.... Imagine to yourself a Man who should love his Watch as we love a Dog, and caress it because he should think himself dearly beloved by it, so as to think when it points out Twelve or One o'Clock, it does it knowingly and out of Tenderness to him."[29] In Wordsworth and even more in Dickens watches really do take on human characteristics, and Bougeant's argument would no longer be effective.

The frequency with which both serious and satiric French works dealing with Cartesian animal mechanism were translated into English is a measure of the great popular interest in the subject. Yet, despite the interest, artists as different as Cowley and Johnson, Pope and Defoe, Swift and Blackmore were able to agree on their opposition to animal mechanism. Only the moderns could be held responsible for the mechanistic philosophy and for the watch analogy that had put order into the fortuitous concourse of Epicurean atoms, but all literary writers—whether ancients or moderns—could agree that animals were more than mere watches. The phenomenon is not easily explained.

One can readily relate the interest in animal mechanism to the popular English editions of Descartes' philosophy that appeared in the years following 1694. Though occasional literary references turn up after *Hudibras* (1663), the greatest concentration occurred in the years 1694–1716; afterwards interest in the subject dwindled rapidly. In *Education* (1693), Locke already describes the Cartesian philosophy as "that which is most in Fashion."[30] The early antagonism to the beast-machine tended to come not from poets at all, but from philosophers and theologians who were worried that this might lead not merely to the man-machine but to the concept that human "cogitation" and even the soul itself operated along mechanical lines.

Despite the risk to their orthodoxy, poets were apparently not prepared to accept a clear-cut dualism between men and animals. They refused to believe that their pets were watches which felt no pain, acted completely without intelligence, and gave their love mechanically. We may note that the reactions of the poets exhibit the quality that was coming to be known as sentiment. The rise of sentiment in literature,[31] and presumably in life, corresponds almost exactly with the period when the poets were reacting to animal mechanism.

The dislike for philosophy and the patent absurdity of equating animals with watches must have contributed to the motivation of the poets, but the sentimental movement should also be taken into account. Indeed, Bougeant actually uses the word *sentiment* to describe what he feels to be the relationship between a dog and a man. The sentimental movement, which began shortly after the scientific and horological revolutions, has never been definitively explained. To the extent that it is a reaction to what has been called "aristocratic cynicism," sentiment may derive in part from the egalitarian tendencies inherent in technological progress and the ever expanding market that it requires and creates.

A series of letters in Defoe's *Review* (1705) may be taken as typical of a considerable number of articles on Cartesian mechanism written by essayists and journalists during the time when the philosophy of Descartes achieved its greatest popularity in England. Fifteen years later, Defoe had a great deal to say in *Moll Flanders* about the craft of taking a watch, but nothing at all about animal mechanism based on the watch analogy. He had a remarkably good instinct for adapting his topics to the time and the medium. As editor of *The Review*, he set up the popular question of animal mechanism through a letter asking "Whether a Dog may not properly be said to

THINK on Things past and to come." Defoe's reply, written from "the Scandal Club," begins: "The Society are not Ignorant, that their Answer to this will be a little wide from the Schoolmens Rules of Philosophy, *That Matter cannot Thin*[*k*]." He lists a number of canine actions (which some people still choose to call intuitive), and remarks perceptively, "If this be not Thinking . . . 'tis a thing every way in its degree equal to that we call Thinking."

Defoe's reply elicited a remarkable letter of some two thousand words from J.S., which appeared in *A Supplement to the Advice from the Scandal Club* (January 1705). The defence by J.S., "upon the Account of Religion," is carefully and seriously argued. The writer is clearly inhibited from equating men with animals on account of the human soul. Like Johnson much later, when Mr. Deane tried to insist on "the future life of brutes,"[32] J.S. cannot allow such feelings as he has for animals to interfere with his orthodoxy.[33]

One is bound to wonder whether a closely argued letter by J.S. written "upon the Account of Religion" in the year following the publication of *A Tale of a Tub* might not be by Swift himself. If it were, it would help to explain the author's point of view regarding the relationship between the soul and the body behind the satiric indirection of the *Tale* and the *Mechanical Operation of the Spirit*. (Curiously this companion piece to the *Tale* was consciously written in a mock "Epistolary Stile.") Like Descartes, J.S. argues "I know I think, and consequently have a Soul, and find no such thing in a Brute; why therefore should I debase Man, the Glory of the Creation, beneath the level of a Beast, for in their Bodies and Senses they excel us; and 'tis the Soul or Thought only makes the incomparable Difference?"[34] The Master Houyhnhnm similarly points to the superior physical endowments of his species, but confesses that in "any Country where the *Yahoos* alone were endowed with Reason, they certainly must be the governing Animal."[35]

The Master Houyhnhnm reported to the grand assembly on the English custom "of *castrating Houyhnhnms* when they were young, in order to render them tame." It cannot be without significance that when Gulliver returns to "civilization," the first money he "laid out was to buy two young Stone-Horses [i.e. not castrated]" with whom he conversed for four hours every day. The Master Houyhnhnm had described Yahoos as men who, "degenerating by Degrees," came to differ from Gulliver, "a *Yahoo*, only a little more civilized by some Tincture of Reason." In England, Gulliver continues "to lament the Brutality of Houyhnhnms," but he will "always treat their Persons with Respect" for the sake of the Master Houyhnhnm and his race, "whom those of ours have the Honour to resemble in all their Lineaments, however their Intellectuals came to degenerate."[36] Clearly Jonathan Swift and J.S. agree on the essential difference between men and horses being the quality of intellect and reason. Moreover, neither will permit any sentiment they may have for animals to inhibit their orthodoxy. Their argument "upon the Account of Religion" still had some weight—as is evident from Defoe's servile reply to J.S., or Addison's circumspection when making animal mechanism the subject of Spectator no. 120—but on the whole Swift and J.S. would appear to be in a minority.

Pope's position seems to be more typical of poets. Though too much may have been made of his fondness for dogs, Pope's lifelong humanitarianism—as expressed in *Guardian* no. 61 and elsewhere—is well known. In Spence's *Anecdotes* Pope deplores his friend Hales' dissection of animals. In the anecdote that follows (no. 269, also dated January 1744) he says: "Man has reason enough to know what [it] is necessary for him to know, and dogs have just that too." Clearly the physical mechanism of the physiologists—like the later mental mechanism of the behavioral psychologists—was generally unsuited to the sensibilities of poets.

John Ray, the naturalist, can speak for the moderns. He attacks Cartesians and Aristotelians alike on the question of giving material souls to animals: "That the Soul of Brutes is material, and the whole Animal, Soul and Body, but a mere Machine, is the Opinion, publickly own'd and declar'd, of *Des Cartes*, *Gassendus* . . . and others. The same is also necessarily consequent upon the Doctrine of the Peripateticks. . . . I should rather think Animals to be endu'd with a lower Degree of Reason." How else might one explain why a dog seeing that a table is too high will first leap on a stool: "If he were a machine, or Piece of Clockwork, and this Motion caus'd by the striking of a Spring, there is no Reason imaginable" why the dog should not attempt to jump on the table, whether or not it is too high. Ray appeals to both the common sense and the sentiment of his readers: "If Beasts were Automata, or Machines, they could have no Sense, or Perception of Pleasure, or Pain and consequently no Cruelty could be exercis'd towards them." But this is contrary to the cries "they make when beaten, or tormented, and contrary to the common Sense of Mankind, all Men naturally pitying them. . . ."[37]

Mandeville, in his *Fable of the Bees*, makes a comparable plea: "When a Creature has given such convincing and undeniable Proofs of the Terrors upon him, and the Pains and Agonies he feels, is there a follower of *Descartes*, so inur'd to Blood, as not to refute, by his Commiseration, the Philosophy of that vain Reasoner?" In the latter part of the seventeenth century Mandeville had himself been a Cartesian, and upheld the thesis "*Bruta non sentiunt.*" In "The Sixth Dialogue," Mandeville puts his changed feelings to good scientific use: "Brutes make several distinct Sounds to express different Passions by: As for Example; Anguish, and great Danger. . . ." Mandeville is a pioneer in insisting that language evolves naturally and does not have a divine origin.[38]

During the eighteenth century, an outright denial of the existence of the soul would have been unacceptable. Put crudely, the choice in considering the souls of men and animals lay between making both souls material (Hobbes), the man's soul incorporeal and the beast's soul material (Descartes), or both souls incorporeal. The first was even more theologically suspect than the last. Since sentiment and observation suggested that animals and men had much in common, there was a tendency to argue that both had incorporeal souls, though of a very different order.

Blackmore—who was, like Mandeville, a physician with poetic aspirations—wrote epics of interminable length that made him one of the chief targets of the Scriblerians. Yet on this subject (despite the difference in style) he is quite close to Pope. He argues that unless animals have a soul, however base, it would be impossible to distinguish "an Aminal from a Watch" since "a Clock of the smallest Size, with wonderful Diversity of minute

Springs and Movements, and great Variety of lasting Motions, does not differ in Kind from a plain one of the largest Dimension, that has but one simple Motion, and that of short Continuance."[39] Blackmore's ornate analogy belongs to one of the patterns of watch metaphors. It may be compared with Boyle's "as if a man should allow, that the laws of mechanism may take place in a town clock, but cannot in a pocket watch,"[40] and with Prior's concept that a watch stripped of "the added movements" still retains the identity of a functional watch so long as the "horal orbit" continues to operate.

Cowley also uses the "common sense" argument against animal mechanism: "What is there among the actions of beasts so illogical and repugnant to reason? When they do any thing which seems to proceed from that which we call reason, we disdain to allow them that perfection, and attribute it only to a natural instinct: and are we not fools, too, by the same kind of instinct."[41] This can be compared to Mat (Matthew Prior in *Alma*) questioning the treatment of horses as mere machines:

> Dick, from these instances and fetches,
> Thou mak'st of horses, clocks and watches.

Prior goes on to insinuate a much greater heresy:

> Quoth Mat, to me thou seem'st to mean,
> That Alma [the mind] is a mere machine.

To this Dick hurriedly responds with a caution that might be applied to all analogies:

> . . . in argument
> Similes are like songs in love:
> They much describe; they nothing prove.

Though the frequency of references declined, the poets' reaction to animal mechanism continued throughout the horological revolution. In *Rambler* no. 41, Johnson is still attempting to differentiate between the reason of men and the instinct of animals. He feels that "memory is the purveyor of reason," and allots to men a greater portion of "memory, which makes so large a part of the excellence of the human soul."[42] Like Swift and J.S., Johnson, "upon the Account of Religion," will not risk giving animals a soul comparable to that of men, but he refuses to rationalize cruelty. Johnson acts as the bridge to a new age. In the *Idler* of 17 June 1758, he sweeps aside Cartesian animal mechanism with much the same common sense and vigor that he applied to other "mechanical" Augustan traditions like the three unities, the chain of being, and the partiality for imitation. Johnson derides in no uncertain terms "the Cartesian, who denies that his horse feels the spur, or that the hare is afraid when the hounds approach her."

In the first half of the eighteenth century (and Hogarth's first and second *Stages of Cruelty* represent this well), people were beginning to question cruelty to animals. Samuel Butler—through the image of "whipp'd Tops and Bandy'd Balls"—may be alluding, as early as 1663, to the manner in which animal mechanism allowed cruelty to be rationalized:

> . . . they now begun
> To spur their living Engines on.
> For as whipp'd Tops and Bandy'd Balls,
> The Learned hold are Animals:
> So Horses they affirm to be
> Mere Engines, made by Geometry. (*Hudibras* 1.2)

Two generations later, Swift set up a speculative situation in which the horses had sense and men were the "Mere Engines." Prior had earlier made the reversal in his *Epitaph on True*, Queen Mary's dog:

> Ye Murmerers, let *True* evince,
> That Men are Beasts, and Dogs have Sence.[43]

And William Somerville, in the highly popular *Chase* (1735), suggests a similar reversal for man made vain by science who treats the "brute creation" as "clock-work all and mere machine."[44]

The simple reversal of Prior and Somerville becomes much

more complex in *Gulliver's Travels.* Swift's four books each offer a particular perspective on animal mechanism, as well as on horology. The Lilliputians, who have no clockwork, think that Gulliver's watch is an animal; the Brobdingnagians, who are expert in clockwork, think that Gulliver is himself "a piece of Clock-work"; the Laputans demonstrate all the absurdity of philosophy and mathematics; and the Houyhnhnms—who will not subdivide time beyond the natural divisions related to the sun and the moon—pose the problem of rational beasts and bestial men.

The following is a description of Gulliver's watch as reported by a Lilliputian:

> Out of the right Fob hung a great Silver Chain, with a wonderful kind of Engine at the Bottom. We directed him to draw out whatever was at the End of that Chain; which appeared to be a Globe, half Silver, and half of some transparent Metal.... He put this Engine to our Ears, which made an incessant Noise like that of a Water-Mill. And we conjecture it is either some unknown Animal, or the God that he worships....[45]

Apart from Swift's insinuation that the watch might be Gulliver's "bourgeois" god, this passage alludes to the tendency for more primitive people to think that watches were a form of animal. Such references are to be found in several works during the early years of the horological revolution. In the *History of Most Manual Arts* (1661), the author maintains that "A King of *China* upon his first seeing a Watch, thought it a living creature, because it moved so regularly of itself, and thought it dead when it was run out, and its pulses did not beat." *Some Years Travels into Divers Parts of Africa and Asia the Great*, which had run into its third edition by 1665, is typical of the voluminous travel literature that Swift was satirizing. It, too, reports on the king of China: "Horologic knowledge they want, as may be supposed by that story of the King who upon first view of a Watch presented by an *European* was so surprized that he verilie believed it a living creature...."[46]

Boyle was also taken with the story. In one place he says that "a clock is not acted by a vital principle, (as those Chineses thought, who took the first, that was brought to them out of *Europe*, for an animal,) but acts as an engine." He uses the story again in a manuscript: "If some rude *Indian* or perhaps one of those Chineses that took Clockes for living Creatures, should perceive the motion of the hand without hearing any other noise than that of the ballance for almost an hour; and if he then heard the Clocke strike twelve, he would suppose that there was some new Accident to this new Animall." The primitive misconception that watches were animals may have been thought to support the idea of animal mechanism.

There is a remarkably parallel quotation in Cicero's *De Natura deorum* 2.34–35, in which it seems to be suggested that if a mechanized globe were taken among savages they would think of it as a living being. The mechanized globe, as Price suggests, developed into our own clocks; the irony of history is that Cicero's British savages also developed into the fashioners of our horological revolution. As Cicero puts it: "But if anyone brought to Scythia or Britain the globe (sphaerum) which our friend Posidonius [of Aphameia, the Stoic philosopher] recently made, in which each revolution produced the same (movements) of the *sun* and moon and *five* wandering stars as is produced in the sky each day and night, who would not doubt that it was by exertion of reason?"[47]

Bolingbroke, writing closer to the time of *Gulliver's Travels*, uses the same basic story, but his emphasis is quite different: "Carry a clock to the wild inhabitants of the Cape of Good Hope. They will soon be convinced that intelligence made it, and none but the most stupid will imagine that this intelligence is in the hand that they see move, and in the wheels that they see turn."[48]

Swift would probably have been glad to agree that the Lilliputians, or British, were "stupid," but he has reserved a different role for the Brobdingnagians. Being expert in clockwork, their technology is as advanced as that of the British. However, the king refuses to control others through the use of gunpowder, and considers Gulliver's "Lilliputian" compatriots the most odious little vermin that ever crawled upon this earth. Although the king was as learned as anyone in the kingdom, "and had been educated in the Study of Philosophy, and particularly Mathematicks; yet when he observed my Shape exactly, and saw me walk erect, before I began to speak, conceived I might be a piece of Clock-work, (which is in that Country arrived to a very great Perfection) contrived by some ingenious Artist. But when he heard my Voice, and found what I delivered to be regular and rational, he could not conceal his Astonishment."

In a neat reversal from the episode in Lilliput, the King naturally enough thinks of Gulliver as a clockwork automaton. He feels at first that Glumdalclitch and her father had "taught me a Sett of Words to make me sell at a higher Price." What finally convinces the King that Gulliver must be more than a clockwork automaton are the rational answers that he continues to be able to give. It is this which makes Gulliver seem "rationis capax," though not necessarily "animal rationale," and it is "upon this great foundation of misanthropy," as Swift says, that "the whole building of my travels is erected" (letter to Pope of 29 September 1725).

The Laputans, like the king of Brobdingnag, are educated in philosophy and mathematics, but they take this study to extremes and are lashed by Swift's satire. The story of the ill-fitting suit of clothes for which the measurements were taken with a quadrant is as good an example of scientific irrationality as any. The reversal in the fourth book concerning animal mechanism (Yahoo) and the rational mind (Houyhnhnms) has already been mentioned. As Laputa is a Lilliput debauched with scientific excesses, so the Master Houyhnhnm represents the quintessence of nature and reason, a Brobdingnag that refuses to countenance even the technology of clockwork.

Clearly the content of Augustan literature was much concerned with clock metaphors and with the clockwork inherent in animal mechanism. The influence of horology on the language and form of Augustan literature was just as great. It will provide the topic of our next chapter.

Part IV

The Reaction of the Romantics

Augustan Clockwork and Romantic Organicism

The Influence of Clocks on Language and Form

The Horological Revolution and Augustan Literature

Since the 1930s, critics like Arthur O. Lovejoy, Marjorie Nicolson, and R. F. Jones have demonstrated the important impact of science on literature during the period with which we are concerned. Some writers have also recognized the relationship between order, regularity, and method in Descartes and similar qualities in French and English classical literature.[1] But there has been no study of the possible relationship between the prime symbol for order and regularity and the period when those qualities were most highly valued in modern literature. In England, the British horological revolution coincides almost exactly with the literature of the Augustan or neoclassical age.[2]

Though the vision is sometimes distorted by their reaction, those who follow an age are frequently best equipped to single out its main features. The Augustans did not think of their predecessors as clockwork poets; the Romantics frequently did. Cowper said that Pope

> Made poetry a mere mechanic art;
> And ev'ry warbler has his tune by heart.

Writing in the shadow of Pope's not inconsiderable reputation, Cowper damns him with faint praise. We need have no doubts about Cowper's views on making "poetry a mere mechanic art." Just over a hundred lines earlier in *Table Talk*, he has given us his views on "clockwork" poetry:

> When labour and when dullness, club in hand,
> Like the two figures [clock Jacks] at St. Dunstan's stand,
> Beating alternately, in measur'd time,
> The clockwork tintinabulum of rhyme,
> Exact and regular the sounds will be;
> But such mere quarter-strokes are not for me.[3]

Here then is a clock simile. It represents (as we have now become accustomed to expect) both order and regularity. What is changing is the attitude of the author towards those qualities.

It became customary to think of the Augustans as mechanical imitators of the classics. A. W. Schlegel, the spokesman for German Romanticism, frequently thinks in these terms, and the clockwork simile very naturally complements such thought. In the celebrated *Vorlesungen über Schöne-Litteratur und Kunst* (1801–1802), while discussing "mechanischen Regeln," he says that "Die Werke mechanischer Kunst sind todt und beschränkt.... So dient z.B. eine Uhr die Zeiten zu messen, weiter kann sie nichts." A hundred pages later, we find him using the same clock simile to represent a rigid mechanical art, incapable of the organic development that vital art requires: "die Kunst soll.... wie die Natur selbständig schaffend, organisirt [*sic*] und organisirend, lebendige Werke bilden, die nicht erst durch einen fremden Mechanismus, wie etwa eine Penduluhr, sondern durch inwohnende Kraft, wie das Sonnensystem, beweglich sind...."[4] Like Cowper, Schlegel thinks of poetry as a creative experience not to be controlled through mechanism, like a pendulum clock.

Tieck—who, in the prologue to *Der Gestiefelte Kater*, mocks a desire for "die Regeln" as middle-class Philistinism—relates literary "rules" directly to the clock in his amusing prologue to *Kaiser Octavianus*:

> Und immer scheut das Dorf die Kosten,
> Das macht die Uhr nun ganz zunichte,

> Denn Werk und alle Räder rosten
> ....
> Die *Einheit* fehlt dem ganzen Werke
> Es läuft nun gegen alle *Regel*,
> ....
> Die *Ordnung* ist nun auch begraben
> Und alles schwimmt in Anarchie, . . . [italics added].[5]

Abrams' *Mirror and the Lamp* is particularly helpful in demonstrating the change from a mechanically to an organically oriented aesthetic: "Since both mechanism and organicism (implicitly asserting that all the universe is like some one element in that universe) claim to include everything in their scope, neither can stop until it has swallowed up the archetype of the other."[6] Abrams is not directly concerned with the metaphor of the clock. But since this was the only effective mechanical metaphor during the neoclassical period, we can readily substitute the clock for his "one element" that "all the universe is like." The clock can also be substituted for Abrams' "mechanical process" in the following quotation: "The basic nature of the shift from psychological criticism in the tradition of Hobbes and Hume to that of Coleridge can, I think, be clarified if we treat it as the result of an analogical substitution—the replacement, that is to say, of a mechanical process by a living plant as the implicit paradigm governing the description of the process and the product of literary invention."[7]

The popularity of the plant analogy coincides with a marked rise in the popularity of botanical research after 1760, just as the popularity of the clock metaphor coincided with the horological revolution.[8]

Romantic Organicism and the Reaction to Clockwork Values

If we were obliged to point to a particular work that signals the change from a mechanical to an organic form of poetry, it would be Edward Young's *Conjectures on Original Composition*, published in 1759. At that time, while attacking the mechanical imitation of his period, Young still had to use the mechanical arts as an example of the progress he desired in the humanities: "Since copies surpass not their *Originals*... while arts mechanic are in perpetual progress, and increase, hence the liberal are in retrogradation and decay." In defining what he means by "an *Original*," Young is clearly aware of the mechanical and botanical analogues for literature between which he stands: "An *Original* may be said to be of a *vegetable* nature; it rises spontaneously from the vital root of genius; it *grows*, it is not *made*: Imitations are often a sort of manufacture wrought up by those *mechanics*, *art*, and *labour*, out of pre-existent materials not their own."[9] Lessing's important "Literaturbrief No. 17," published in the same year, is well worth comparing. It lauds Shakespeare, questions the "Rules," and berates the French Corneille for his "mechanical" imitation of the ancients.

We can never fully understand a period without considering its relationship with what came before. Just as Erasmus Darwin drew his theory of evolution from an analogy with Hartley's mechanical learning process through association, so Coleridge developed his organic imagination from the same concepts of mental mechanism. Coleridge had at one time been sufficiently impressed with Hartley to call his son after him. But in the *Biographia*, when reviewing and criticizing the history of mental mechanism culminating in Hartley, Coleridge makes it clear that his faculties of memory and fancy rather than imagination, are intended to incorporate everything that is valid in the mechanical eighteenth-century theory of association.[10] Thus, in Coleridge's organic theory of imagination, the mechanical processes of memory and fancy ("a mode of memory") are incorporated but downgraded. They provide, in Coleridge's words, the materials out of which poetry can grow: "In association then consists the whole mechanism of the reproduction of impressions in the Aristotelian psychology. It is the universal law of the passive fancy and mechanical memory; that which supplies to all other faculties their objects, to all thought the elements of its materials."[11] For the organic vital quality of the *imagination*, Coleridge feels obliged to coin the new word *esemplastic*; almost immediately, he couples this to the names *Linnaeus* and *Darwin*, with whose biological studies his reader could readily identify. The imagination "is essentially *vital*, even as all objects (as objects) are essentially fixed and dead,"[12]

Ironically, though he downgraded the mechanical aspects of man, Coleridge may have been inspired to use the most Romantic of all analogies through a Cartesian mechanistic concept. It has become a commonplace of criticism that the Aeolian wind harp is the analogy most typical of Romantic poets, one without which they might have lacked a conceptual model for the way that the creative mind and imagination work. Coleridge's *Dejection: An Ode* has been cited as providing the "earliest inclusive instance of this symbolic equation."[13] But an earlier draft, related to Coleridge's *Eolian Harp*, suggests an important relationship between the so-called Romantic metaphor and Cartesian mechanistic thought:

> And what if all of animated Life
> Be but as Instruments diversely fram'd
> That tremble into thought, while thro' them breathes
> One infinite and intellectual Breeze
> ....
> Thus God would be the universal Soul,
> Mechaniz'd matter as th' organic Harps
> And each one's Tunes be that, which each calls I.[14]

One and a half centuries before, in his *De la Formation du foetus*, Descartes had said that to explain "la Machine de nostre Corps... nous en avons de juger qu'il y a une Ame dans une horloge, qui fait qu'elle monstre les heures."[15] The early Coleridge, as we have just seen, still employed the concept of a "universal Soul" inside "Mechanized matter." Both the organic imagination and the Aeolian harp incorporated mechanical conceptions originally developed with the essential aid of the clock metaphor.

But for the Romantics clockwork analogies were generally pejorative. One could continue to duplicate examples of the Romantic reaction to the mechanical model. Cowper castigates schoolmasters who, concentrating on syntax, "Dismiss their cares when they dismiss their flock—/Machines themselves and govern'd by a clock"; he is almost as harsh with conversation that, instead of flowing gently, seems "as if rais'd by mere mechanic powers."[16] Wordsworth feels that the *Cave of Staffa*, home of the Ghost of Fingal, is a "fit school/For the presumptuous thoughts that would assign/Mechanic laws to agency divine." In *The Excursion*, he tells us: "Moral truth/Is no mechanic structure, built by rule;... / ... but ... like the water-lily, lives and thrives."

Cowper, too, had noted the superiority of nature's clockwork. He addresses the "Yardley Oak":

> By thee I might correct, erroneous oft,
> The clock of history, facts and events
> Timing more punctual, unrecorded facts
> Recov'ring, and misstated setting right—[17]

Perhaps the biographer of Turner, in the nineteenth century, may speak for the age. After praising Turner's painting, he says that "Turner was a dumb poet.... though ... sometimes a happy epithet offers a sustained clock-beam cadence imitative of Pope."[18]

How the Augustans Felt about Their Mechanical Rules for Art

It is clear that the Romantics frequently thought of the Augustans as producers of a clockworklike mechanical poetry whose form and content were stultified by rules and imitation. When we come to examine the Augustans, we find that the question is more complex. The Augustans themselves, for example, occasionally complain about mechanical, clockworklike writing. Dryden's position is ambivalent. In the epilogue to *An Evening's Love*, he makes a critic say of the drama:

> here's so great a stir
> With a son of a whore farce that's regular,
> A rule, where nothing must decorum shock!
> Damme, 'tis as dull, as dining by the clock.[19]

Yet we know that ten years later Dryden and his age were remarkably proud of the way that he put Shakespeare's *Antony and Cleopatra* into the neoclassical straitjacket of *All for Love*.

Generally, the real Augustans—while ostensibly approving Pope's "RULES" and "*Nature Methodiz'd*"[20]—thought only of poetasters as purveyors of mechanical writing. Pope's project in *Peri Bathous*, for producing poetry, like clocks, by batch production has already been mentioned. In *Dunciad* book 4, his schoolmasters are almost as mechanical and insensitive as Cowper's:

> As Fancy opens the quick springs of Sense
> We ply the Memory, we load the brain,
> Bind rebel Wit, and double chain on chain, . . .[21]

Even Hogarth, whose art is so much involved with clocks, feels that clockwork movement is antithetical to beauty. In the *Analysis of Beauty*, he uses the example of Vaucanson's famous duck—"a little clock-work machine, with a duck's head and legs fixt to it"—to demonstrate how much more graceful are the movements "of nature's machines (one of which is man)" than "those made by mortal hands."[22]

The impact of horology on the literature of the Augustan period is complicated by the fact that it influenced both the literary ancients and the scientific moderns. As the handmaiden of astronomy and the leading element in the technology of the age, it was connected with the important movement to simplify and regularize the language in the interests of science. Furthermore—in an age when scientific questioning coincided with the weakening of religious and feudal bonds—horology repaid its debt to astronomy be providing such essential metaphors for order as the Cartesian clockwork animal and the Newtonian clockwork universe. In addition the "fixing" of the language by Johnson's *Dictionary* (1755) marked the conclusion of a century during which spelling, syntax, and punctuation became relatively mechanized and therefore "modern."

Thus far, the Augustans parallel the moderns. Whatever the underlying contradictions in Augustanism, their ideal was simplicity in language and decorum in action. But neoclassicists responded negatively to the scientific concept of "progress" that developed in the seventeenth century and continues to be with us. Unlike the moderns, they still felt or claimed that the world was in decay, that their forefathers had been giants, and that the best they could do was to imitate the ancients. Ironically, it was in imitating the ancients that Augustans employed the clockwork mechanical rules for which they have been denounced.

Like clockwork, whose analogy he so frequently used, Descartes influenced both the moderns and the ancients. On the one hand, his ideas of infinity and progress were modern and proto-Romantic; on the other hand, he had a profound effect on classical concepts of language and method.

Marjorie Nicolson has illustrated admirably the dichotomy in the Cartesian influence which has bedevilled attempts to place Descartes squarely in the one camp or the other. She suggests that "few . . . seem to have noticed the effect of the Cartesian idea of indefinite extension upon one of the most significant of all seventeenth-century conceptions: the idea of infinity. . . . In this conception lies the key to the characteristic form taken in England . . . by the idea of progress, and to one of the most profound changes which occurred in seventeenth-century thought."[23]

Naturally, this "Romantic" aspect of Descartes is stressed by Swift. In the *Battle of the Books*, Aristotle attacks the two great moderns, Bacon and Descartes. He misses Swift's countryman, but, having "found a *Defect* in his *Headpiece*," hits Descartes. As a result, Death draws Descartes "into his own *Vortex*." The vortices of Descartes appeared to involve a "fortuitous" element in the heavens quite at variance with the clockwork metaphor with which Cartesian mechanism is associated.

Swift (using those very terms) valued "Reason" and "Memory" over "Enthusiasm" and "Imagination." In section 8 of the *Tale of a Tub*, he explains "Inspiration" and "Aeolists" in a manner remarkably predictive of the Aeolian wind harp as a Romantic metaphor. But for Swift, of course, this is a prelude to section 9, "A Digression concerning Madness," which prophetically undercuts many of the future values of the Romantics.

The issue is complicated by the fact that whatever Swift may have thought of Cartesian and other philosophers, his own language, like that of his age, had been radically affected by

Descartes in particular and the scientific moderns in general. It is to them that we must credit the decisive change towards modernity in English prose style during the last half of the seventeenth century. Not merely Addison, Swift, and Defoe, but even Dryden, Boyle, and Bishop Sprat, writing in the third quarter of the seventeenth century, no longer seem to live in the same world as had Robert Burton and Sir Thomas Browne.[24]

Descartes and the moderns were responsible not only for changing the language the Augustans used, but the order and regularity that Descartes and the mechanistic philosophers imposed on mental conceptions with the aid of the clock metaphor even more directly affected the form of Augustan composition. The demand for order and regularity went far beyond regularizing syntax and eliminating figures of speech; during the horological revolution punctuation, word meanings, and even spelling also became mechanically regularized. Marjorie Nicolson stresses the "effect so-called Cartesianism had upon *method* in literary composition. Certainly the insistence that the process of thought has a logic like that of mathematics, that it is characterized by *order* and *regularity* . . . that its touchstones are *clarity* and *lucidity*—all this was not without its effect upon English literature" (last four italics added).[25]

In the Augustan period, then, the moderns had overcome the old science and its prophet Aristotle, had demonstrated that the world was not in decay, had insisted that knowledge lay in practical experiment rather than dead classical tomes, and had convinced large sections of society that science and technology could provide the foundation for progress. But the Augustans represented a reactionary element in society. With Swift they insisted that though Aristotle the scientist may have acknowledged that he "proceeded in many things upon Conjecture,"[26] Aristotle the representative of Greek classical literature had a very different status. His *Poetics* provided a set of rules derived from what was demonstrably the finest literature in the history of man; the Augustans could not only satirize some of the excesses of the scientists, but they could claim that man as a social and literary animal was in decline. Hence the underlying pessimism of a Pope who questioned whether man was properly "plac'd" on the "great chain" of being, and a Swift for whom man's nature seemed fated to decline in the direction of the Yahoo. Like Hogarth's *Tailpiece*, their world ends in the pessimism and chaos of the last book of *The Dunciad*, when Time himself dies and his daughter "*Truth* to her old cavern fled."

But the movement represented by Young and the later Romantics rejects order and mechanism for the new concepts of progress and growth. In the second half of the eighteenth century, Dr. Johnson—who has been called too late and too great to be an Augustan—symbolically provided the *coup de grâce* to three of the most important Augustan tenets. His *Dictionary* (1755) stabilized the language that the Augustans had claimed to be in decline; his review of Soame Jenyns' *Free Enquiry* (1757) seriously questioned the order and hierarchy of the chain of being; and his *Preface to Shakespeare* (1765) challenged the unities of time, action, and place with which French and English classicists had been attempting to shackle modern drama.

The Clock as a Symbol of Order and Reason during the Horological Revolution

Part of the purpose of this book has been to trace the use of the clock as the seventeenth century's main analogy for order both in organic life and in the universe. The same science which questioned the old order of the universe set up the demand both for an accurate clock to prosecute its research and an effective analogy through which to counteract the potential for chaos that such research seemed to forbode in society. A line from Donne illustrates vividly a current fear that the old world had ceased to exist: "'Tis all in peeces, all cohaerence gone. . . ." The quotation is from Donne's *Anatomie of the World*, "wherin, By occasion of the untimely death of Mistris Elizabeth Drury, the frailty and decay of this whole World is represented." Donne complains that order has ceased to exist not only in "the Planets, and the Firmament," but also in the chain of being where "Prince, Subject, Father, Sonne, are things forgot."[27]

Yet if the organic world of Donne's "First Anniversary" is "all in peeces," the "Funerall Elegie," that immediately follows, points very clearly to the new mechanistic analogy for order:

> But must wee say she's dead? may't not be said
> That as a sundred clocke is peecemeale laid,
> Not to be lost, but by the makers hand
> Repollish'd, without errour then to stand, . . .[28]

This is not an atypical example. Herbert, Descartes, and Clauberg also explain the difference between a living and a dead body by comparing them respectively to a watch in running order and a watch that has stopped.

Pope's "Hail, Bards triumphant, born in happier days" may be cited as one of the many illustrations proving that the Augustans felt the world to be in decay, even though the scientists now thought otherwise. Indeed the concept of decay may help to explain the neoclassical predilection for imitation of the ancients rather than for original invention.[29] In addition, of course, imitation is a more mechanical process which coincides both with the social requirements of the time and with the mechanistic philosophy. Descartes, despite his circumspection, seems to suggest, at the end of *L'Homme*, the possibility of mental processes being comparable to the mechanical operation of a watch.[30]

Imbued as we now are with Romantic values, it may be difficult for us to appreciate that Dryden intends no disparagement, in his *Preface to the Fables*, when introducing Ovid and Chaucer as imitators: "Both writ with wonderful facility and clearness; neither were great inventors: for Ovid only copied the Grecian fables, and most of Chaucer's stories were taken from his Italian contemporaries, or their predecessors." When we learn that "Troilus and Cressida" was "much amplified by our English translator, as well as beautified," we are reminded of Pope's "What oft was *Thought*, but ne'er so well *Exprest*." Dryden goes on to defend Chaucer's imitation by a topical analogy with manufacturing, which might well refer to the pendulum and balance spring inventions of Huygens that he had seen so profitably exploited by the British: "The genius of our countreymen, in general, being rather to improve an invention than to invent themselves, as is evident not only in our poetry, but in many of our manufactures."[31] Every period seems to commend great artists for aspects of their work most sympathetic to the age.

Just as imitation was an aspect of composition sympathetic to mechanical rules, the pervasive desire for reason and order also coincided with such rules. Long before neoclassicism, Aquinas had used an analogy with clocks to demonstrate order and reason: "All the things moved by reason display the order of reason, though they themselves are without reason, for instance.... The same appears in the movement of clocks and other works of human art."[32] After claiming that heathen poets (the early divines) were less involved in zeal, discord, and controversy than modern divines, Hobbes condemns any Christian poet "reasonless" enough to wish to "be thought to speak by inspiration, like a bagpipe."

At this point, Hobbes produces his well-known formula for the mechanical operation of the "poet's" mind: "Time and education beget experience: experience begets memory; memory begets judgement and fancy; judgement begets the strength and structure, and fancy begets the ornaments of a poem. The ancients therefore fabled not absurdly, in making Memory the mother of the Muses." Memory provides us with the world "as in a looking-glass... whereby the fancy, when any work of art is to be performed, finds her materials at hand and prepared for use."[33] The mechanical system of Hobbes appears to leave no room for a creative imagination which might result in works not controlled by the memory. Through his downgrading of imagination, as "nothing but *decaying sense*," and his upgrading of the memory, he helped to promote the hegemony of the latter in English aesthetics for more than a hundred years.

The call for reason and order in poetry during that century—which coincides with the British horological revolution—is too well known to require many examples. Dryden's defence of the "regularity" of English drama (in his role as Neander in the *Essay of Dramatic Poesy*) is curiously ambivalent. The British never accepted the mechanical rules for drama as wholeheartedly as the French, but they continued to pay a curious lip service to them throughout the Augustan age.

Dryden can also write tongue-in-cheek about the rules. The speaker of the "Epilogue" to *Aurengzebe* says of its author:

> A pretty task! and so I told the fool,
> Who needs would undertake to please by rule.

But others laid down more precise instructions. Dillon, Earl of Roscommon, has the following to say in the heroic couplets of his *Essay on Translated Verse*:

> On *sure Foundations* let your *Fabrick Rise*,
> And with attractive *Majesty* surprise,
> Not by affected, *meretricious Arts*,
> But strict *harmonious Symetry* of *Parts*.

The heroic couplet, typical of the Augustan age, was a fine mechanism for Pope, but it also offered a regular and ordered clockwork admirably suited to the talents of less competent poets.

In drama, too, critics recognized the need for a mechanical framework. Rymer, in *Tragedies of the Last Age*, points to "the *proportions*, the *unities* and *outward* regularities" as "the *mechanical part* of Tragedies."[34] At the turn of the century, Barker, in *The Polite Gentleman* (1700), speaks for many in his age: "With the good leave of our *Poets*, the whole Secret of their Art consists in finding Terms, and ranging them in good Order."[35]

One cannot stress too much the fact that some poets and critics during the Augustan age always remained well aware of the need for imagination, or what Nietzsche would later call the Dionysian element in art. But more than at any time during the history of English poetry they were frequently also aware of the need to control the Dionysian by a strong Apollonian form.

Gerard, in his *Essay on Genius* (1774), offers a balanced view during the time of change. He allows us to savor the mechanical and the clockwork connotations that had become associated with order and the rules. Gerard has the following to say of the balance needed between what can now be termed the Apollonian and the Dionysian: "Neither the fertility nor regularity of imagination will form a good genius, if the one be disjoined from the other. If fertility be wanting, the correctest imagination will be confined within narrow bounds,... If regularity be absent, an exuberant invention will lose itself in a wilderness of its own creation."[36] Gerard does not expect genius to operate "like a mere machine," because the poet must select the subject, and "is continually employed in choosing the proper tracts of thought." But, on the other hand, he is equally opposed to an imagination that will "form a confused chaos, in which inconsistent conceptions are often mixt, conceptions so unsuitable and disproportioned, that they can no more be combined into one regular work, than a number of wheels taken from different watches, can be united into one machine."[37]

Summary: The Influence of Clocks on Language and Form

By way of summary, the idea of progress came first to scientists and later to poets. In order to counteract the potential for social and political chaos inherent in scientific progress, scientists, philosophers, theologians, and poets alike sought a new symbol for order and regularity. At the same time, astronomers, who were in the vanguard of science, needed more accurate clocks to forward their work. The revolution in horology was a direct outcome of the astronomers' demands, and the more accurate watches and clocks that it produced provided the necessary symbol for order, so long as this was required.

The mechanistic philosophy, from Descartes onwards, systematized the new technologically oriented form of order: both the universe and organic matter operated like clocks. But theologians had reservations regarding the relationship of the soul with mechanical men, while poets—as we have observed—were remarkably antagonistic to the Cartesian concept of clockwork animals.

Nevertheless, the general acceptance of the idea that the new heliocentric universe operated on a mechanical model led readily into the concept of God as watchmaker which became widely accepted during the horological revolution. But the overthrow of the Aristotle who represented ancient science did not lead immediately to the overthrow of the Aristotle who represented such giants of the past as Homer, Aeschylus, Sophocles, and Euripides. Despite contrary trends, the Augustans adhered to their claim that literary and social men continued to decline.

During the horological revolution, the mechanical imitation and rules of neoclassicism, like the general demand for a "Mathematical plainness" in language, were admirably suited to the clockwork pattern for order, reason, and clarity also being adopted. But when attention became focused on the biological sciences, from about 1760, the clockwork model for scientists, philosophers, and poets proved less adaptable to related concepts of growth and evolution.

In aesthetics, the change from a mechanical to an organic model is marked by Young's *Conjectures on Original Composition* (1759). In philosophy, Hume and Kant moved in a comparable direction. Hume questioned and Kant finally cast out—at least for philosophers—the teleological argument for the existence of God. Furthermore, the "crude *Egoismus*" from the *Wissenschaftslehre* of Kant's student Fichte encouraged a new race of giant poets. Imbued with *romantische Ironie*, the new race, to use Coleridge's words, found "its spring and principle within itself."[38] For Romantic poets, the "imitation" of Aristotle's *Poetics*, and the clockwork rules of the Augustans were an anathema, and degrading.

Young's *Conjectures* mark the change in poetry both from a mechanized to an organic analogue, and from an Augustan belief in decline to a Romantic belief in growth and progress. Thus, insofar as the belief in progress is concerned, the Augustan poets, at least, were a century behind the scientists. After the eighteenth century, only theologians continued to press the argument from design based on the example of a watch. It is perhaps one of the greater ironies in history that the clock analogy—at first so radical and so hard to reconcile with orthodox views—now survived through Paley as a respectable if uninspiring retainer in the ranks of piety. The argument from design continued to serve theology until Darwin's *Origin of Species* (1859) all but gave it the *coup de grâce*.

The Ambivalence of the Novelists

The English Nineteenth-Century Novelist: An Ambivalent Attitude towards Clocks and Clockwork Values

Although Romantic poets almost always tended to react adversely to the mechanical qualities in life and art suggested by clockwork regularity, the reaction of the novelists was by no means as antagonistic. Despite what Sterne may have thought of Walter Shandy, novelists, for the most part, did not disparage clockwork regularity until after the early nineteenth century. And even then the reaction of a novelist like Dickens was frequently ambivalent.

In the century before Dickens, novelists tend to favor the order inherent in clockwork. The actions of Richardson's incomparable Sir Charles Grandison seem to be regulated by the watch that he so frequently holds in his hand. Earlier, Richardson's Pamela had fulsomely praised Mr. B. for the clockwork regularity that he showed in his life. Boswell, in the *London Journal*, attempts to achieve the ideal of a comparable sense of order. It is true that much of the attraction of the *Journal* derives from a Boswell who seems to imagine himself a latter-day version of the lady-killer Macheath. But we sense, before long, that he will ultimately enter the law and the relative "reserve and dignity of behaviour" that this imposes.[1]

At the very time when he is confined to his rooms, by venereal disease contracted through the actress Louisa, Boswell rejoices: "My present life... very fortunately is become agreeable. My affairs are conducted with greatest regularity and exactness. I move like clock-work" (9 February 1763). Later in the *London Journal*, though not under the same circumstances, Sir James Macdonald speaks highly of the life at Oxford because "time was regularly laid out. Exactly at such hours he did such and such things, the doing of which in that manner was his pleasure, and could scarcely be interrupted, as he moved like clock-work" (30 March 1763). Boswell approves highly, "I really believe the college life in England is the least painful of any."

Goethe makes a comparable case for the value of clockwork regularity. In *Wilhelm Meister*—after the death of Mignon, who represents art—the hero goes on his travels, and Goethe describes the value of clocks in the practical and pedagogical sphere of life: "Time [is] the highest gift of God and Nature, and the most assiduous handmaid of existence. Clocks have been multiplied amongst us, and one and all indicate the quarters with hand and stroke." Goethe, as a scientist in his own right, is an early advocate of the effective transmission of time: "telegraphs... give... the course of the hours by day and night." The author understands all too well that there is an ethic related to the well-ordered life: "Our moral theory... is furthered in the highest degree by division of time and attention to every hour. Something must be done at every moment, and how could this be effected if attention were not paid to the work as well as to the time."[2]

In *Fortunes of Nigel* (1822), Scott goes beyond Goethe by even marrying his aristocratic hero to the beautiful daughter of the watchmaker David Ramsay. Though Scott "romanticizes" his life, Ramsay actually was the horologer to James I, as he is in the novel. The real Ramsay kept his shop near Temple Bar, and was the first master of the Clockmakers Company. Much of the technical material that Scott uses has been taken from Derham's *Artificial Clock-Maker*.[3]

Scott treats Ramsay rather like an absentminded professor, who is involved with abstruse mathematical and horological problems when he should be attending to his customers.

Margaret Ramsay's godfather, Master George, tries to bring his old friend to his senses by a good-natured use of the clock metaphor: "Lay by these tablets, or you will crack the inner machinery of *your* skull, as your friend yonder has got the outer-case of his damaged." James I decides that his fellow Scotsman really is of aristocratic descent: "David Ramsay is no mechanic, but follows a liberal art, which approacheth almost to the act of creating a living being." Here, as frequently, the clock, the automaton, and the mechanical metaphor are not far apart. Scott makes James I say of his clockmaker: "We propose to grant him an augmented coat-of-arms, being his paternal coat, charged with the crown wheel of a watch in chief, for a difference; and we purpose to add Time and Eternity, for supporters, as soon as the Garter King-at-Arms shall be able to devise how Eternity is to be represented."[4]

In the second half of the nineteenth century, novelists like Lewis Carroll, Dickens, and Hardy, while still involved with horological imagery, demonstrate a much greater ambivalence towards it than had Richardson, Goethe, or Scott before them. Carroll's ambivalence, however, differs somewhat from that of Dickens and Hardy. Carroll subscribes to the Romantic myth regarding the spontaneity of creative art. Yet if we examine Carroll's novels in the light of their carefully structured horological imagery it becomes hard to avoid the conclusion that the "spontaneous" creation must have benefitted from some "mechanical" revision.

Lewis Carroll (Charles Lutwidge Dodgson) was one of the most methodical men who has ever lived. Between 1861 and 1898 he made a précis of every letter he wrote or received. The last entry in the cross-reference system under which he listed these is 98,721. Carroll was also "a clever mechanist" with musical boxes and similar automata. Though one may suspect that he "doth protest too much," he endorses the traditional post-Romantic opposition to clock-oriented writing. He says, "I cannot set invention going like a clock, by any voluntary winding up; nor do I believe that any *original* writing . . . was ever so produced."[5] Many of Carroll's "single ideas" probably did come as inspiration, but his finished work has surely been consciously structured by certain patterns relating to clocks and time.

Though there are other remarkably regular patterns in the Alice books our concern is with those related to clocks and time. The two works are of almost indentical length, and each is divided into twelve parts. In terms of calendar time they divide exactly the twelve months of the year. *Alice in Wonderland* takes place on 4 May, Alice Liddell's birthday, and *Through the Looking Glass* on 4 November, exactly six months later when the real Alice is seven and a half years old. *Alice* is a tale of the outdoors, croquet, and the spring; *Looking Glass* is a tale of the indoors, chess, and the autumn.

Although for obvious reasons the time sequence cannot be explicitly stated, there is an implicit suggestion that *Alice*'s twelve parts cover the hours of daylight, whereas the twelve parts of *Looking Glass* pass through the hours of night. Alice starts with the making of a daisy-chain (Chaucer's "day's eye") on a hot day, and concludes with Alice's sister "watching the setting sun." *Looking Glass* starts late in the November day. Dinah, the Liddell's cat, had finished washing the black kitten "earlier in the afternoon." But in the last paragraph of *Looking Glass*, Alice says to the black kitten (who had presumably spent the night as the wife of the Red King): "As if Dinah hadn't washed you this morning!" The Red King is the character in whose consciousness Alice herself may exist ("you're only one of the things in his dream") during the night; she had taken the place of his wife during her remarkable game of chess. Perhaps Carroll's illusions extended beyond the looking glass.

Time is involved in the content as well as the structure of the Alice books. Clocks have an important part at the beginning of both. In *Alice*, Alice, at first, found "nothing so *very* remarkable about the White Rabbit . . . but, when the Rabbit actually *took a watch out of its waistcoat-pocket*, and looked at it, and then hurried on, Alice started to her feet." Structurally, this is what Gustav Freytag in *Technique des Dramas* calles the *erregende Moment*.

At an equally important point, the *peripeteia* or *Wendepunkt*, Carroll introduces two other memorable creatures, the Mad Hatter and the March Hare. They are the most clock-afflicted of all Carroll's characters. The Mad Hatter's watch, as we learn at some length during the conversation at the tea party,[6] is always set at the right time for six o'clock tea. The Mad Hatter is said to portray Theophilus Carter, an Oxford furniture dealer who invented an alarm clock bed that tipped its occupant onto the floor at the preselected hour of his choice. Again at the tea party, after a passage dealing with the elasticity of "Time," the Hatter describes the occasion when he was singing "Twinkle, twinkle, little bat," and the "Queen bawled out 'He's murdering the time! Off with his head!' . . . And ever since that . . . he [time] won't do a thing I ask! It is always six o'clock now."

In *Looking Glass*, the *erregende Moment* also occurs in part 1. When Alice is kneeling on the mantlepiece beside the clock, her fantasy—"Let's pretend the glass has gone all soft like gauze"—turns into reality. One of the first things she notices in "Looking-Glass land" is that "the very clock on the chimney-piece (you know you can only see the back of it in the Looking-Glass) had got the face of a little old man, and grinned at her." Two hundred years after the beginning of the horological revolution, Alice's mechanistic analogy relating the clock to a little old man was a harmless enough metaphor. It belonged to the realm of fancy and fantasy, rather than *L'Homme Machine* and the atheism with which that had been associated. Finally—and surely beyond the coincidence attributable to a completely unaided creative imagination—the Hare and the Hatter (thinly disguised as Haigh and Hatta in *Looking Glass*) make their appearance in part 7 of both works.

By way of contrast, the ambivalence of Dickens seems to differentiate between good and bad horological images. In *Great Expectations*, he is well aware of the sinister nature of Estella's mechanical heart or the expensive gold repeater in the hands of Mr. Jaggers. For Dickens, however, a large, old, and well-used watch like Sam Weller's evokes some of the nostalgia that the Romantic poets reserved for sundials. His novels, as we have earlier noted, are permeated with clock references. The author himself seems frequently to have compared clocks to human beings. In a letter to John Bennett (14 September 1863), Dickens complains that since his hall clock was returned from cleaning "it has struck the hours with great reluctance and, after enduring internal agonies of a most distressing nature, it has now ceased striking altogether. Though a happy release for the clock, this is not convenient to the household."[7]

An author who frequently employs the pathetic fallacy slips easily into the sentiment associated with *Master Humphrey's Clock*. When the old friends meet in fraternal comfort, Master Humphrey's old clock not only contains but almost embodies "as many papers as will furnish forth [their] evening's entertainment." In a passage "From his Clock-Side," Master Humphrey writes: the friends have "lingered so long over the leaves . . . that as I consigned them to their former resting place, the hand of my trusty clock pointed to twelve, and there came towards us upon the wind the voice of the deep and distant bell of St. Paul's as it struck the hour of midnight." This permits the author to describe at length a visit to the old clock of St. Paul's, constructed by Langley Bradley: "Its very pulse, if I may use the word, was like no other clock . . . . as if its business were to crush the seconds as they came trooping on, and remorselessly to clear a path before the Day of Judgment." There is the potential here for portraying a diabolic element in clockwork, but that is not Dickens' intention. He continues, "the fancy came upon me that this was London's Heart, and that when it should cease to beat, the City would be no more. . . . Does not this Heart of London, that nothing moves, nor stops, nor quickens,—that goes on the same let what will be done,—does it not express the City's character well?"[8]

There is another group, respected members of a lower class, who also enjoy listening to stories in Master Humphrey's house; this group is presided over by Mr. Weller, or rather by "Mr. Weller's Watch" from which it takes its name. The passage introducing the watch is one of the best loved in Dickens:

> Unbuttoning the three lower buttons of his waistcoat and pausing for a moment to enjoy the easy flow of breath consequent upon this process, he laid violent hands upon his watch-chain, and slowly and with extreme difficulty drew from his fob an immense double-cased silver watch, which brought the lining of the pocket with it, and was not to be disentangled but by great exertions and an amazing redness of face.[9]

At one point, Master Humphrey's group was informed that "Mr. Weller's Watch" had adjourned its sittings from the kitchen, and "regularly met outside our door . . . for the convenience of listening to our stories." They were allowed in, given chairs at a respectful distance, and "the clock wound up, we entered on our new story." In this case, the story was *Barnaby Rudge*.

When Dickens at last brought the weekly magazine to a close, Master Humphrey passed away sitting peacefully by his clock. In a short concluding passage, the Deaf Gentleman describes the sentimental circumstances under which the remaining friends, who include Mr. Pickwick, gather to read the will that the trusty clock has been found to contain. Dickens closes with the lines: "Our happy hour of meeting strikes no more; the chimney-corner has grown cold; and MASTER HUMPHREY'S CLOCK has stopped for ever."

Thus the work that gave the world such Victorian favorites as *The Old Curiosity Shop* and *Barnaby Rudge* did so within the framework of two groups of storytellers, not to mention a whole city, all of which were represented by their respective horological symbols. There is certainly the potential for diabolism in Dickens' clocks—the bell of St. Paul's warns of the Day of Judgement, and repeaters are significant in the characterization of Scrooge and Mr. Jaggers—but the daemonic, like the Gothic, is softened by sentiment.

We have previously noted in Dickens the apparent dichotomy between the practical writer, "in all things as punctual as the clock at the Horse Guards," and the nostalgic sentimentalist who has a taste for old clocks. Dickens' attitude towards repeaters seems also to involve a certain ambivalence. In the hands of Scrooge at midnight there must surely be something ominous about them. On this point a comparison of Pope with Dickens is symbolic of the respective ages that they represent. Pope's Belinda wakes at midday; in a gesture of typical rococo pride, she presses her repeater when she can well enough see the dial. Dickens' Scrooge is sitting up at midnight. A child of his age, he is so conscious of time that, when the hour rings out from the neighboring church, he presses the repeater "to correct this most preposterous clock." We shall find that there are many clock-oriented slaves in Poe's bourgeois borough of Vondervotteimittiss—if we need to go so far in order to discover people who are continually checking on the time.

Fifteen years after publishing the *Christmas Carol* (and in return for reading that well-loved work), Dickens was himself presented at Coventry with "a gold repeater of special construction by the watchmakers of the town; as to which he faithfully kept his pledge to the givers, that it would be thenceforward the inseparable companion of his workings and wanderings, and reckon off the future labours of his days until he should have done with the measurement of time." Forster, who wrote this account of Dickens' watch, was bequeathed it on the author's death. He left the repeater to a mutual friend, Carlyle, in whose family is still remains.[10]

David Copperfield and Pip, of *Great Expectations*, are the two main characters fashioned on the life of their author; in the latter novel, we find what surely is the golden repeater that Dickens had recently received. The older Dickens of *Great Expectations* (1860–61) seems more concerned than formerly with his art. Pip oscillates between the simple values of Joe and Biddy and the heartless beauty that is his Estella, or star—perhaps the art to which he aspires. In attempting to gain the approbation of Estella, Pip tries to become a gentleman. Unbeknown to him the funds for achieving this social metamorphosis are supplied by Magwitch, a wealthy convict.

Estella, for her part, lives in the house of the rich Miss Havisham. All of that lady's clocks have stopped at twenty minutes to nine, the moment when long ago she was jilted by a man. Estella's indoctrination turns her into a beautiful automaton trained to break men's hearts by way of retribution for what happened to Miss Havisham. Thus *Great Expectations* is the tale of two selfish adults who try to fulfill through the indoctrination of the young what they could not achieve themselves. Indeed, Pip says of his early experience in Miss Havisham's home, "I only suffered in Satis House as a convenience... a model with a mechanical heart to practice on when no other practice was at hand."[11]

The only character who seems to know all the facts and even to control the others as puppets is the sinister Mr. Jaggers. As a lawyer, he represents both Miss Havisham, who is Estella's guardian, and Magwitch, who is Estella's true father. He is the guardian of Pip, and he has complete control over the services of

Estella's real mother. Moreover, he wears and is indeed symbolized by the watch which is surely the gold repeater with its massive watch-chain that the author, Dickens, had received some two years earlier (chap. 25).

But there are forces for good operating on the edge of (and sometimes within) the circle, centered around Jaggers, of those who seek to make automata of Pip and Estella. Pip is saved by the *caritas* of Joe and Biddy, and he, in his turn, goes some way towards softening the mechanical heart of Estella. In *Hard Times*, Dickens juxtaposes the heartless mechanical values of Coketown (symbolized by Gradgrind's "deadly statistical clock") with the sentiment of Cecilia Jupe and the circus people.

In *Far from the Madding Crowd* (1874) Thomas Hardy continues the ambivalent attitude towards clocks and watches that we find in Dickens. In this early novel of Hardy the horological images have for the first time been moved away from the normal urban setting and superimposed on the seasonal pattern of country life in a Wessex hamlet. Before we look at the two most antithetical images—the watches of Gabriel Oak and Sergeant Troy—let us consider how many of the characters are associated with appropriate watches and watch images. Boldwood—as might be expected of a man with "Roman features"—has "a time-piece, surmounted by a spread eagle";[12] Cain Ball's large watch dangles in front of him in comic emulation of the large old watch belonging to Gabriel whom he serves; and Old Coggan has inherited an "old pinchbeck repeater," atypically benevolent because it is made of a cheap alloy of copper and zinc. Coggan's repeater seems to strike by itself at three crucial stages of the story (pp. 242, 274, 314–15). Though poor Fanny Robin has no watch, Hardy underlines her predicament by using the pathetic fallacy to describe the chimes of neighboring clocks at three crucial stages in her story (outside Troy's barracks, on the way to Casterbridge Union-house, and when her corpse awaits its transfer back to Weatherbury).

Hardy not only suggests a pathetic fallacy in our association with clocks but he is also like Dickens in indicating the nature of his protagonists through their watches. Nothing could be more nostalgic and benevolent than Gabriel's watch, which is several years older than his grandfather. It is lovingly described at length in the first chapter of *Far from the Madding Crowd*, as one that "may be called a small silver clock . . . a watch as to shape and intention, and a small clock as to size." Readers would surely compare this watch with that of Sam Weller. Gabriel's large old silver watch is clearly meant to provide a sharp contrast with the aristocratic gold watch that is the only heirloom of the diabolical Sergeant Troy. Troy's watch, too, is central to the story: it epitomizes the qualities that Bathsheba finds attractive in Troy in the "Hay-mead"; it contains Fanny's blond hair, "which has been the fuse to this great explosion"; and it remains with Bathsheba after Troy is apparently dead.

Cut down to its bare essentials, *Far from the Madding Crowd* involves a quadrangular relationship. Bathsheba must choose between three men: Boldwood with a clock "surmounted by a spread eagle," Gabriel with his benevolent watch as large as a silver clock, and Troy who not only has a watch with malevolent connotations but who also refuses to marry Fanny after waiting for her under the diabolical quarter jack, the automaton in the All Saints Church of chapter 16. That chapter was inserted by Hardy as a horologically significant afterthought. But perhaps even more significant is the single and singular reference in chapter 15 to the timepieces with which Bathsheba chooses to replace the previous possessions of her uncle. What she buys for her mantlepiece are "great watches, getting on to the size of clocks." At least in terms of the horological imagery, it would seem clear that Bathsheba's tastes lean towards Gabriel's long before she is prepared for a marriage that will cement the fate of their two thousand contiguous Wessex acres.

The ambivalence of such nineteenth-century novelists as Dickens and Hardy, whose horological images are far from being entirely malevolent, is not easily explained, particularly when one considers how uniformly antagonistic were such contemporary poets and short story writers as Baudelaire, E. T. A. Hoffmann, and Poe. Very possibly an important contributory cause lay in the demands that novel writing made on the author. Dickens, Trollope, and Hardy produced a considerable volume of material. Much of it first appeared in serial form, the very nature of which emphasized a need for time-oriented production methods. In recent years, the techniques of organization and methods study have demonstrated how weak were the arguments of clerical employees in maintaining that, unlike factory workers, they could not be subjected to time measurement. The novelists might have been just as vulnerable. Their private knowledge that successful novel writing requires a regular and time-oriented application to work may not be unrelated to their concern with clocks and their tendency to be less uniformly antagonistic towards them than were the Romantic poets.

It so happens that Trollope's posthumously published *An Autobiography* (1883) deals with the application of time study to novel writing. Some readers have unfortunately felt that either Trollope's actions or their revelation was an offense to the dignity of literature.[13] Copernicus and Galileo had to await the impunity of death before publishing the great myth-questioning works which gave impetus to the horological revolution—Trollope's posthumous work is neither as profound nor as iconoclastic, but the Romantic myth which it questions is far from being exploded. Trollope says, "I wrote my allotted number of pages every day. . . . And as a page is an ambiguous term, my page has been made to contain 250 words." He kept a journal of his production—very much like that which a work-study engineer would keep today. In order to produce the more than seventeen novels and much else that appeared from 1859 to 1870, he became to slave to his watch: "It was my practice to be at my table every morning at 5.30 A.M. and it was also my practice to allow myself no mercy. . . . It had at this time become my custom . . . to write with my watch before me, and to require from myself 250 words every quarter of an hour. I have found that the 250 words have been forthcoming as regularly as my watch went."[14]

Trollope was not alone. Dickens, too, was bound to the wheels of time. We have already referred to his claim that he was "in all things as punctual as the clock at the Horse Guards."[15] George Sand, despite her other activities, wrote regularly from midnight until 4:00 A.M., and her collected works fill more than one hundred volumes. Her detractors relish the story that when, on one occasion, she had finished a novel at 1:00 A.M. she promptly started another. There is clearly a dichotomy between the Romantic conception that art should derive from a "spontaneous overflow of powerful feelings," and the novelists' strongly felt need for method. The relationship between clock time and the writing methods of nineteenth-century novelists warrants further research.

Writers of the American Renaissance: Differing Trends in the Attitude Towards Technology

Despite the fact that the climate of literary opinion seems to have become increasingly antimechanical in the period after the horological revolution, the British novelists, as we have indicated, are generally ambivalent in their attitude toward clocks. Certainly they are rarely as passionately opposed to the clock as the poets and "dark Romantics" with whom we shall be concerned in the final chapter. Like their British contemporaries, American novelists of the nineteenth century were rather ambivalent in their attitudes towards technology, but certain different trends are worth noting.

When Tompion died in 1713, he had produced the remarkable quantity of about 6,000 watches and 550 clocks. Tompion's achievement marked the transition from the individual *manu*facture and design of a watch to batch production, the division of labor, the relative interchangeability of parts, and the use of machines for such operations as wheel and screw cutting. In the first half of the nineteenth century, the Americans took the next decisive step warranted by the exponential growth in watch and clock manufacture. Five years after he introduced the shelf clock in 1817, Eli Terry (probably influenced by Eli Whitney's mass-production methods for muskets) was making 6,000 a year at $15 each. In 1842, Chauncey Jerome began exporting brass clocks to England for $1.50 each. Whitney himself turned to clockmaking after 1848 in partnership with Aaron Dennison and Samuel Curtis. Ironically, the American concept of making watches by machine led to the subsequent Swiss dominance of the market.

Though technology influenced literature on both sides of the Atlantic, there were differences of emphasis. By the first half of the nineteenth century, the British had seen the rise and fall of

Tompion travelling clock, ca. 1700. This fine clock—which includes alarm, striking, and repeating mechanisms—formerly had arrangements whereby it could be controlled by either a pendulum or a balance wheel. Tompion made about six thousand watches and five hundred clocks during his lifetime. (*Lent to Science Museum, London, by Mrs. M. L. Gifford.*)

Eli Terry, Plymouth Ct., Shelf Clock, ca. 1816. These pillar and scroll clocks were adapted to the techniques of mass production and heralded a lucrative new era in the clock industry. Five years after he introduced the shelf clock, Eli Terry was making six thousand a year at fifteen dollars each. (*By courtesy of Yale University Art Gallery, Bequest of Olive L. Dann.*)

their horological industry. They had also developed the use of steam in three crucial stages: for pumping water out of mines (Savery 1698, Newcomen 1705, Watt 1763 and 1769); for contributing to the mechanical power of factories in the industrial revolution (Watt 1781, though water power was more important in the early stages); and for producing the revolution in transport during the first half of the nineteenth century. Long association with technological development had made influential British writers such as Carlyle, Dickens, Ruskin, Morris, Butler, Wells, and Kipling very much aware of the diabolical potential of machinery.[16]

In America in the second quarter of the nineteenth century, there was a remarkable upsurge in the number of steam-powered factories, steamships, and, above all, railroads. In 1830, seventy-three miles of railway track had been laid in America; by 1860, there were more than thirty thousand miles. The American writer, like the British, tended to regret the daemonic and mechanical elements of steam power and horology. But he was also an American, aware that something different and exhilarating was occurring in his society. Like Abraham Cowley and the Dryden of *Annus Mirabilis* or *To Charleton* one and a half centuries earlier, American Renaissance writers (those of the period 1829–70) sometimes felt of their technological revolution that "Bliss was it in that dawn to be alive." For example, Walt Whitman says, in the lengthy paean that is *Passage to India*:

> A worship new I sing,
> You captains, voyagers, explorers, yours,
> You engineers, you architects, machinists, yours.
> You, not for trade or transportation only,
> But in God's name, and for their sake O soul.

Whitman said of *Passage to India*, "There's more of me, the essential ultimate me, in that than in any of the poems."[17] Is it a coincidence that the very word "technology" was coined by Jacob Bigelow, the Harvard professor, in 1829, the first year of the American literary Renaissance?

Technological progress also had its dark side. Leo Marx writes in *The Machine in the Garden*: "The ominous sounds of machines, like the sound of the steamboat bearing down on the raft or of the train breaking in upon the idyll at Walden, reverberate endlessly in our literature. . . . it is difficult to think of a major American writer upon whom the image of the machine's sudden appearance in the landscape has not exercised its fascination."[18] However, the very passage in Hawthorne's notebooks that Marx uses to demonstrate the conflict between locomotive and landscape is inconsistent in its attitude towards machines. Before the disruptive locomotive intrudes upon his musings, Hawthorne includes the striking of the village clock among the peaceful sounds of rural life:

> Now we hear the striking of the village-clock, distant, but yet so near that each stroke is distinctly impressed upon the air. This is a sound that does not disturb the repose of the scene, it does not break our sabbath; for like a sabbath seems this place, and the more so on account of the cornfield rustling at our feet. It tells of human labour, but being so solitary now, it seems as if it were on account of the sacredness of the Sabbath. Yet it is not so, for we hear at a distance, mowers whetting their scythes; but these sounds of labour, when at a proper remoteness, do but increase the quiet of one, who lies at his ease, all in a mist of his own musings.[19]

The clock seems to have been so thoroughly incorporated into daily life and into Hawthorne's basic ways of thinking about human civilization and society that he cannot repudiate it, any more than he can the mowers' scythes and the cornfield. References to clocks and technology in Thoreau, and, in a more complex manner, in Melville show this same ambivalence. While aware of the harmful potential of technology, frequently they accept or even welcome the machine into their pastoral dream.

The railroad is omnipresent in Thoreau's *Walden* (1854). It passes on the other side of the pond from the house, which Thoreau (to escape "the fretful stir unprofitable") has built from the dismantled timbers of a railway worker's hut. In Thoreau's "pastoral," one is never more than two or three pages from an incident or symbol connected with the "Fitchburg Railroad": "I usually go to the village along its causeway and am, as it were, related to society by this link."[20]

Likewise, references to clocks and clock time occur throughout *Walden.* Thoreau frequently states his opposition to lives regulated by clock time. Men, who "have become the tools of their tools," "have no time to be anything but a machine"; they "must learn to reawaken . . . not by mechanical aids, but by an infinite expectation of the dawn." For the natural man "the day is a perpetual morning. It matters not what the clocks say or the attitudes and labors of men." At Walden, "my days were not the days of the week, bearing the stamp of any heathen deity, nor were they minced into hours and fretted by the ticking of a clock." However, mechanical symbolism enters readily into Thoreau's imagery for man: his "mainspring is vanity, assisted by love of garlic and bread and butter." Not even nature is exempted from the clock metaphor, for the whippoorwills "begin to sing almost with as much precision as a clock"; and the red squirrel "would be in the top of a young pitch-pine, winding up his clock."[21] Thus it is not only the railroad that brings technological civilization to Walden woods—Thoreau's own imagination introduces it there.[22] Thoreau must draw on his own experience to describe imaginatively the natural inhabitants of the woods, and, in nineteenth-century New England, that experience could not but be influenced by the pervasive presence of the clock.

At other times, Thoreau gives a new dimension to his treatment of clock imagery, one that reflects, albeit not without irony, the euphoria of most nineteenth-century Americans about technology. In 1884, the railway was to add a new dimension to time measurement through the international "zones" and standard time advocated by Sandford Fleming, Canadian surveyor and engineer of the C.P.R. In *Walden*, Thoreau is already aware of how the railway would add a new dimension to the mechanical control of man in society. Like the telegraph which served it, the railway synchronized time: "They go and come with such regularity and precision, and their whistle can be heard so far that farmers set their clocks by them, and thus one well conducted institution regulates a whole country. Have not men improved somewhat in regularity since the railroad was invented?"[23]

In Melville's imagery, time regulation and the railway also

have their place. In *Moby-Dick*, during the final three days' chase of the white whale, the whalers rely on a clockworklike quality of the whale—its need to surface at regular intervals to breathe. On every occasion that the whale went below, "at the well-known methodic intervals," Ahab, "binnacle-watch in hand," "would take the time" and order that he be lifted to his perch "so soon as the last second of the allotted hour expired." Melville compares the whale's remarkable regularity of timing to "the mighty iron Leviathan of the modern railway... so familiarly known in its every pace that, with watches in their hands, men time his rate as doctors that of a baby's pulse." However, in Melville's universe the clockwork regularity of natural order vies with an irrational power moving behind things. Speaking of the whaler's skill in timing the whale, Melville draws this example: "But to render this acuteness [in timing] at all successful in the end, the wind and the sea must be the whaleman's allies; for of what present avail to the becalmed or windbound mariner is the skill that assures him he is exactly ninety-three leagues and a quarter from the port?" In fact one of the omens presaging the tragic end of Ahab's quest is the failure of his very first attempt to time the white whale. Seeing Moby Dick sound on the first day of the chase, "'An hour,' said Ahab, standing rooted in his boat's stern." However, it is only a few minutes later that the whale appears, rising from below to attack Ahab's boat.[24]

Melville is well aware that the interplay of irrational daemonic force and rational mechanistic regularity is a human as well as a natural phenomenon. The two forces—one mechanical, the other the natural energy of the physical world—appear in two images Ahab uses, in immediate succession, to describe his success in instilling the crew with his own mad passion for revenge: "'Twas not so hard a task. I thought to find one stubborn, at the least; but my one cogged circle fits into all their various wheels, and they revolve. Or if you will, like so many ant-hills of powder, they all stand before me; and I their match." Ahab opposes his will to both the unreflecting animal regularity of the crew and to the first mate Starbuck's appeal to reason and the sane, domestic order which is expected to regulate human life. The latter's impulse is specifically compared to clockwork when Starbuck, confronted by Ahab's lust for revenge, finds "my soul is more than matched; she's overmanned; and by a madman! Insufferable sting, that sanity should ground arms on such a field! But he drilled deep down, and blasted all my reason out of me.... my whole clock's run down; my heart the all-controlling weight, I have no key to lift again."[25]

Reacting to *Ethan Brand* in a letter to Hawthorne (1? June 1851) Melville employed a new version of the image of the watchmaker God: "The reason the mass of men fear God, and *at bottom dislike* Him, is because they rather distrust His heart, and fancy Him all brain, like a watch."[26] Melville's most fully developed clock analogy—that in the "Chronometricals and Horologicals" pamphlet in *Pierre*—is quite like the eighteenth-century image of the watchmaker God and the divine clockwork of his universe. However, for Melville this image, rather than guaranteeing order, ensures confusion. In Melville's analogy, certain particularly sensitive human souls are like chronometers, adjusted to the time that comes from God, "the great Greenwich hill and tower from which the universal meridians are far out into infinity reckoned."

> Now in an artificial world like ours, the soul of man is further removed from its God and the Heavenly Truth, than the chronometer carried to China, is from Greenwich. And, as that chronometer, if at all accurate, will pronounce it to be 12 o'clock high-noon, when the China local watches say, perhaps, it is 12 o'clock midnight; so the chronometric soul, if in this world true to its great Greenwich in the other, will always, in its so-called intuitions of right and wrong, be contradicting the mere local standards and watch-maker's brains of this earth.

However, Melville concludes that "in this remote Chinese world of ours" to carry Greenwich time—i.e., to follow heavenly truth—is local folly, which God cannot have intended for common men since "such a thing were unprofitable for them here and, indeed, a falsification of Himself, inasmuch as in that case, China time would be identical with Greenwich time, which would make Greenwich time wrong." Even if God is "all brains like a watch," men cannot know what time it is—they can only know that their own time is, in one way or another, wrong.[27]

Looking back to the heyday of the Cartesian mechanistic philosophy in the seventeenth century, and the time when the God of the horological revolution ruled almost unquestioned as a mighty watchmaker, one can see that, by Melville's time, the overall climate of literary opinion had greatly changed.

The Clockwork Devil
A Romantic Reaction to Clockwork and Clockwork Automata

Poetic Dislike for an Enslavement to Increasingly Accurate Devices for Measuring Time

As we have noted earlier in the chapter on "Augustan Clockwork and Romantic Organicism," the Romantic stress on creative imagination is itself in part a reaction to the horological revolution and the clockwork rules. It is remarkable how few poets attacked the clock and what it represented before the end of the horological revolution. One exception to this rule would seem to derive from poets who did not wish to be controlled personally be the demands of time. There is a surprisingly early poem expressing just such a mood by Dafydd ap Gwilym, the Welsh bard whose poetry is thought to have been written between about 1340 and 1380. "A curse on its weights, a curse on its wheel," he says of the clock which has just woken him.[1]

Rabelais is an author about whom more is known, and his antipathy to clocks is understandable. This study has perhaps not stressed sufficiently the extent to which monasteries were involved in the transition from the bell to the clock (the contribution of monks to early manufacturing in other areas has also been greater than is generally recognized). A man of Rabelais' disposition might tolerate a clock-oriented education for Gargantua, but he knew enough about time-oriented monastic lives to ban clocks from the Utopian Abbaye de Thélème. In that monastery, the only rule was "Fais ce que tu voudras": "And because in all other monasteries and nunneries all is compassed, limited, and regulated by hours, it was decreed that in this new structure there should be neither clock nor dial... for, said Gargantua, the greatest loss of time that I know, is to count the hours."[2]

Though sundials sound no bell, they evidently could exert a comparable tyranny. Plautus (c. 254–184 B.C.) has the gods confound that man:

> Who in this place set up a sun-dial,
> To cut and hack my days so wretchedly
> Into small pieces!

Plautus continues, "When I was a small boy,/My belly was my sun-dial."[3] Here is another form of the complaint against horological tyranny. In a later part of his work, Rabelais quotes Plautus on this when saying that for most people "their appetite and their belly was their clock" (4.64). Early in the eighteenth century, when Richmore complains that the time is "almost One," Farquhar's Young Woud'be, in *The Twin Rivals*, makes a comparable point: "Then blame the Clockmakers, they made it so; the Sun has neither the Fore nor Afternoon—Prithee, What have we to do with Time? Can't we let it alone as Nature made it? Can't a Man Eat when he's Hungry, go to Bed when he's Sleepy, Rise when he Wakes, Dress when he pleases, without the Confinement of Hours to enslave him?" But Locke protests that some "have set their stomachs by a constant Usage, like Larms [alarm clocks], to call on them for four or five" meals.[4] The image was common enough. In *English Proverbs*, John Ray includes: "Your belly chimes, it's time to go to dinner."

Another form taken by the reaction to horology involved a Romantic nostalgia that differentiates between clocks and non-mechanical timepieces like sundials and sandglasses. This is coming closer to the Romantic concept of clock as devil, but the very nostalgia demands a muted tone. Cowper—who had attacked the Augustan "clockwork tintinabulum of rhime"—gives as his example of corruption, "What should be and what was an hour-glass once,/Becomes a dice box" (*Task* 4.220–21).

Blake takes the point much further in *Jerusalem*. He juxtaposes a "pastoral" hourglass to the new clock that is symbolic of his dark Satanic mills in Albion or England:

> And all the Arts of Life they changd into the Arts of Death in Albion.
> The hour-glass contemnd because its simple workmanship
> Was like the workmanship of the plowman, & the water wheel,
> That raises water into cisterns: broken & burnd with fire:
> Because its workmanship was like the workmanship of the shepherd.
> And in their stead, intricate wheels invented, wheel without wheel:
> To perplex youth in their outgoings, & to bind to labours in Albion
> Of day & night the myriads of eternity that they may grind
> And polish brass & iron hour after hour. . . .[5]

One can read into what Blake says here, as elsewhere, a surprising awareness of the dangers inherent in technology. Samuel Butler would make the warning more clearly in the middle of the nineteenth century, but we are barely beginning to face the implications of his *Erewhon* even in our own time.

In *On a Sundial*, Hazlitt says, "I never had a watch nor any other mode of keeping time in my possession, nor ever wish to learn how time goes." (Rousseau, the son of a watchmaker, is said to have symbolically discarded his watch upon abandoning Geneva.) Lamb is very definite about his preference for sundials over clocks. Though there is much similarity in the nostalgic tone of Lamb's *Elia* and Dickens' *Master Humphrey's Clock*, the later work has become nostalgic about the very clocks that the earlier work condemns. In the essay "Old Benchers of the Inner Temple," Lamb certainly sees his sundial through a Romantic haze: "It was the primitive clock, the horologe of the first world. Adam could scarce have missed it in paradise."[6] Lamb's nostalgia for sundials involves the image that they have been man's friend all the way back to the time of the Garden of Eden. But we recollect that Plautus' call for the paradise before sundials is comparable to Lamb's call for the paradise before clocks.

Another way in which poets frequently questioned the value of clock time was to suggest that time is an elastic commodity not to be measured in equal gradations. Dryden's *Hind and the Panther* cites examples from the Old Testament:

> For which two proofs in sacred story lay,
> Of Ahaz dial, and of Joshua's day,

Shakespeare implies the elasticity of time in humorous vein when Falstaff claims that he "fought a long hour by Shrewsbury clock" (*I Henry IV* 5.4.151–52).

Since the elasticity of time would seem to be a favorite theme of poets, we shall restrict ourselves to one or two brief examples from each of the centuries with which we are involved. In his *New Inn*, Ben Jonson's Lady Frampul cries out: "O, for an engine to keep back all clocks."[7] In Jonson's *Staple of News*, Pennyboy Junior sets his watch on the table, as it strikes the hour when he comes into his fortune and says to it: "Thy pulse hath beat enough. Now sleep and rest"(1.1). In the seventeenth century, Prior gives a new twist to the old poetic conceit:

> That Cloe may be serv'd in state,
> The hours must at her toilet wait;
> Whilst all the reasoning fools below
> Wonder their watches go so slow.

At the close of the horological revolution, Sterne's *Tristram Shandy* provides, of course, a *tour de force* in ridiculing chronology and clock time. For the nineteenth century, we cannot do better than give the last word to a man so remarkably knowledgeable on the subject of time as the Mad Hatter. As he says to Alice, Time, if properly treated, will do anything to please her. "For instance, suppose it were nine o'clock in the morning, just time to begin lessons: You'd only have to whisper a hint to Time, and round goes the clock in a twinkling! Half-past one, time for dinner!"[8]

We have noted thus far how poets, even without attacking time directly, can suggest that it is not the measure of all things. When they do attack timepieces it is often to claim that sundials and hourglasses are better than clocks, or that the belly is a better measure of time than any of these. But poets also differentiated between particular types of clock.

George Colman, the Younger's *Inkle and Yarico*, performed in 1787, takes an eighteenth-century view in saying that "truth is a golden repeater," a watch that "sets a man right in the dark." The golden repeater was a status symbol throughout the eighteenth and nineteenth centuries. Mr. B. (himself as regular as clockwork) eventually presented Pamela with his mother's "fine repeating watch," after she had successfully negotiated for marriage rather than seduction. But Wordsworth, on behalf of the Romantics, questioned the time and luxury-oriented values of the new middle class. In *The Cuckoo Clock*, he insists that one should "Forbear to covet a Repeater's Stroke," and then goes on to endow the cuckoo clock with the same "pastoral" quality that Lamb has given to the sundial.

Romantic Poets: The Order Inherent in a Clockwork Urban Society as a Characteristic of the Devil

But Romantic poets did not only dislike the material values represented by an expensive watch. They sensed a more immediate threat from the order inherent in the clockwork urban society that surrounded them. The same clockwork which during the horological revolution had symbolized the order needed by theologians, philosophers, and poets alike was now becoming, for some poets, a symbol of the devil rather than of God. Though this is particularly true of the dark Romantics—like Hoffmann, Poe, and Baudelaire—they are far from being alone.

In "Proverbs of Hell," Blake tells us: "The hours of folly are measur'd by the clock, but of wisdom: no clock can measure." There are many images of tyrannic wheels and cogs in Blake's *I Saw a Monk*, *Jerusalem*, *Four Zoas*, and *Milton*. In *Jerusalem*, we are told that "cogs tyrannic" are not to be found in paradise. The negative value Blake places upon horology is evident when in arguing against Swedenbourg he says that "Heaven would upon this plan be but a Clock." There is clearly a reversal here from the values of the mechanistic philosophy, and those frequently expressed during the horological revolution. In this, Blake is typically Romantic.

The fact that poets had come to associate clockwork with the devil is implicit in Coleridge's use of the clock simile, "Like clock-work of the Devil's making," in *Delinquent Travellers*. A century earlier, in his description of a firework display, Addison says, "Within this hollow was Vulcan's shop full of fire, and clock-work" (Guardian no. 103). But he does not make the same allusions to the devilish nature of clockwork that Hoffmann or Poe might have drawn from such a scene.

In Baudelaire, the clock is often a warning of evil, a *memento mori* in a rather special nineteenth-century sense. Its metal voice which speaks all languages warns not only of the evil in death or the life hereafter, but of the ever tighter tyranny that time was imposing on living men:

> Horloge! dieu sinistre, effrayant, impassible,
> Dont le doigt nous menace et nous dit: "*Souviens-toi*!"

Antoine Adam perceptively points to Gautier's comparable statement of man's unequal battle in his *L'Horloge*: "Un combat inégal contre un lutteur caché."[9]

In Baudelaire's *L'Imprévu*, the clock again warns of damnation: "L'Horloge à son tour, dit à voix basse: 'Il est mûr. . . .'" In *Le Voyage*, Baudelaire points out that it matters little whether like some we travel or like others we stay in the same place. Either way the enemy, Time, "mettra le pied sur notre échine." His demand that "nous embarquerons sur la mer des Ténèbres" brooks no refusal. Death is the last voyage, and in Baudelaire's poetry it is a voyage that seems to be little mellowed by Christian optimism.

"O douleur! ô douleur! Le Temps mange la vie," Baudelaire cries out in *L'Ennemi*. And what he says emphasizes the Saturnine and Satanic aspects of Time which refer back to the iconography of an earlier age. At midday or midnight, the voice of the clock is equally gloomy and foreboding. In *Rêve parisien*,

> La pendule aux accents funèbres
> Sonnait brutalement midi, . . .

But in *L'Examen de minuit*, the clock striking midnight makes us question our use of the day which has fled. It is the fate of man, as in *Madrigal triste*, to be "convulsant quand l'heure tinte." Paley's work might still be promoting a watchmaker God, but some poets at least were concurrently warning man about a clockwork devil.

Tennyson's *Devil and the Lady* was published posthumously, but apparently written before 1825. In 1.5, the Devil in soliloquy approaches the timepiece; what he says goes even beyond Baudelaire:

> There is a clock in Pandemonium,
> Hard by the burning throne of my Great Grandsire,
> The slow vibrations of whose pendulum,
> With click-clack alternation to and fro,
> Sound "EVER, NEVER!" thro' the courts of Hell,
> Piercing the wrung ears of the damned that writhe
> Upon their beds of flame, . . .[10]

There had been no clock in Milton's Pandemonium. Nor is there one in *Inferno*; Dante's "horlogue" rings out its joyous bells in Paradise.

The Clockwork Automaton in the Eighteenth Century

We have observed before that the history of clocks and automata shows that they are closely related. Clocks themselves may have derived from an attempt to illustrate or imitate the movements of the universe; automata—in such forms as Jacks, crowing cocks or holy figures—have been associated with clocks from their earliest times. The very nature of the horolgical revolution was to make clocks more and more an article of utility, specializing in such particular areas as the repeater, the chronometer, the bracket clock, the Act of Parliament clock, or the plain pocket watch. The popular eighteenth-century orrery was the automated illustration of the new heliocentric universe. Except in specialized and often relatively simplified situations like the cuckoo clock, there came to be less and less call in horology for the incorporation of automata.

Yet this was by no means the end of automata. Some of the finest models were made in the eighteenth century, frequently for the profitable Chinese and Turkish trade. Since Greek and Roman antiquity, there has been much literary allusion to automata;[11] in more recent years, Albertus Magnus, Regiomontanus, and even Descartes were reputed to have produced such artifacts. Price's "Automata and the Origins of Mechanism," and Bedini's "Role of Automata in the History of Technology" both make a strong case for the work of clockmakers in developing automata being an essential contribution to the scientific and industrial revolutions.[12]

Probably the best known automata of the eighteenth century were Vaucanson's two musicians, and his duck that was credited with "eating, drinking, macerating the Food, and voiding Excrements." He began exhibiting in 1737, and—having become both wealthy and famous—sold his models in 1743. Vaucanson was also the inventor of many important industrial processes including metal cutting machinery and an apparatus for automatic weaving. His automata were both illustrated and described in a pamphlet of some twenty-four pages "sold by *Mr. Stephen Varillon* at the *Long Room* at the *Opera House* in the *Hay-market*, where these Mechanical Figures are to be seen at 1, 2, 5, and 7, *o'Clock* in the Afternoon. 1742."[13] Later in the century, Cox's museum of clockwork wonders became a fashionable "must" for Londoners that is described by Fanny Burney in *Evelina* (1778).[14]

In 1774, the Jaquet-Droz, father and son, founded a firm in London. Bedini describes their three most famous clockwork androids, which have survived: "The Writer, a life-size and lifelike figure of a boy seated at a desk, is capable of writing any message up to 40 letters. The Artist is a similar figure of a boy that makes four sketches.... The third figure is that of a young girl that plays the clavichord by the pressure of her own fingers."

Their fourth piece, the Grotto, seems to have been broken up, under the accusation of witchcraft, when it was sent to Spain.[15]

Vaucanson and the Jaquet-Droz firm specialized in life-sized figures. Though it is difficult to make a direct comparison with earlier automata, it is clear that the later mechanics had a much more sophisticated technology at their service. The careful descriptions of their work would seem also to indicate a closer attention to the realistic reproduction of both anatomy and movements.

Man's dream of creating men in nonmechanical terms would seem to have an even longer history as a literary and imaginative theme. In one variant of the theme a simulacrum or statue comes magically to life (early legends of Vulcan or Pygmalian, the mediaeval Golem of Jewish folklore, the Faustus legend, the affair of Don Juan's father-in-law, and many miraculous animations of holy images); in another variant, pseudo-science supplies a substitute for magic (the alchemical hatching of the homunculus of Paracelsus, or the electric vital fluid supplied by lightning for the monster Frankenstein).[16] It will be clear from the names mentioned that the Romantics took over this tradition in its nonmechanical as well as its mechanical forms.

The plausibility of the tradition had been strengthened during the eighteenth century by the mechanistic philosophy of Descartes, the psychological determinism of Hartley, and the appearance of remarkably versatile modern clockwork androids which lent credence to the age old power dream that man might imitate or even eliminate the function of God. The Romantics seem to have been fascinated and yet horrified by such dreams. Not only were these aspirations evil in the Biblical terms of creating graven images and usurping the power of God, but for the Romantics Cartesianism, psychological determinism, and clockwork were by their very nature anathema. In *Queen Mab*, Shelley expresses the Romantic aversion for automatism because obedience, which makes man a "mechanized automaton," is the "Bane of all genius, virtue, freedom, truth." In the *Letter to Maria Gisborne*, he talks of himself as a "mightly mechanist" preparing to breathe "a soul into the iron heart/Of some machine portentious," and as a "weird Archimage" who is

"Plotting . . . devilish enginery,/The self-impelling steam wheels of the mind. . . ."

Spenser's Archimago has become Shelley's mechanist devil. The female equivalent of Archimago is the evil enchantress—such as Spenser's Duessa or Acrasia, Tasso's Armide, or Gryphius' ghost of Olympia—symbolizing, in Chaucer's words, "Fylthe over-ystrawed with floures." Sometimes the Romantics retain her more traditional role: this occurs with Coleridge's Geraldine; Keats' "belle dame sans merci" and Lamia; Eichendorff's Venus, Diana, and Romana; or the Adelheid of Goethe's *Urgötz*. But on other occasions the new factors transmute the traditional archetype. Hoffmann's Olympia is a clockwork evil enchantress; Dickens' Estella is a psychologically indoctrinated evil enchantress; and Brentano's "schöne Kunstfigur" seems to mix the whole thing up by combining clockwork with fairy-tale magic.[17]

In 1860, Charles Barbara wrote his little known Hoffmannesque tale in which a Major Whittington chooses to live with his own automatons. His mother, his wife, and his daughter were automatons created by himself. This perhaps takes the familial relationship with automatons as far as it will go. But the question of man's relationship with his mechanical productions remained an important one. Eighteenth-century automatons, like those of Vaucanson and the firm of Jaquet-Droz, in addition to mechanical inventions, like Watt's steam engine, gave rise to a very real fear—reflected in such Western works of art as E. T. A. Hoffmann's *Sandman*, Mary Shelley's *Frankenstein*, and Samuel Butler's *Erewhon*—that man might become the slave of his inventions.[18]

Hoffmann's *Sandman*—better known to many through Offenbach's version of the *Tales*—deals with Nathanael's love for Olympia, a beautiful automaton. After rejecting the very human Klara with the words "You damned lifeless automaton," the hero falls hopelessly in love with a real automaton. She is the "daughter" of Spalazani, his physics professor who is apparently involved in making automatons. But Olympia is also claimed by Coppelius (*Coppo* is the Italian for "eye socket"), the sinister maker of eyes and glasses. In a traumatic scene, Nathanael overhears the two men fighting for possession of his beloved:

"Let go! . . . Monster! Villain! Risking body and soul for it? . . . That wasn't our arrangement! I, I made the eyes! I made the clockwork! Damned idiot, you and your damned clockwork! Dog of a clockmaker! . . ."[19]

In other works—such as *Don Juan*, the mechanical movements of Cardillac in Mademoiselle de Scuderi, or the puppets of *The Doubles*—Hoffmann demonstrates his interest in the juxtaposition of man and robot. Nathanael's choice between Klara and Olympia is therefore not untypical of Hoffmann. The mechanical symbolism of Olympia allows us to see in her our modern "evil enchantress"; she is the technology that may lead us to destruction. But what she symbolizes is also as old as the choice made by Gryphius' Cardenio, when, with unholy desire, he follows the ghost in the form of another Olympia. One kiss turns the "Lust-Garten" into a wilderness, and Olympia cries out "dein Lohn die Frucht der Sünde."[20] Men have always sought both sexual power and the opportunity to eat of the tree of knowledge good and evil. But it would seem that since the seventeenth century the fruit of the tree has seduced Western man along paths of technology that were only dreams for his predecessors.

Edgar Allan Poe: A "Dark Romantic" Fascinated by the Clockwork Values That He Attacks

Poe is another of the dark Romantics who is fascinated by the very clockwork that he attacks. *Maelzel's Chess Player* illustrates the fascination in an unusual way. Poe's remarkably incisive arguments for proving that the chess player is not in fact the mechanical automaton that it is claimed to be, demonstrate equally well his careful consideration of the characteristics of clockwork automatons. Elsewhere, his Sheherezade undercuts man's supposed mechanical genius,[21] and *The Colloquy of Monos and Una* demonstrates that Poe had concepts of true "*duration*" that stood apart from "the irregularities of the clock upon the mantel."

As occurs with other Romantic poets, Poe's real abhorrence for clockwork derives from the extent to which he feels that it is enslaving man. His poem *The Bells* suggests, in its four stanzas, a pattern of development through life. The bells, as symbols of time, become more and more menacing. From youth to death they are "Keeping time, time, time,/In a sort of Runic rhyme." The first stanza deals with the "Silver bells" of "a world of merriment," and the second with "mellow wedding bells,/Golden bells." This concerns the first part of life when the enslavement of time seems relatively less ominous and Dylan Thomas could say, "I sang in my chains like the sea." But in the final two stanzas Poe's bells become quite different. At first, they are "alarum bells—/Brazen bells/ . . . In the startled ear of night." By the last stanza, the warning of the clock has become much like that of Baudelaire. The "Iron bells" are now ghouls, whose king tolls, keeping time "To the moaning and the groaning of the bells."

The *Devil in the Belfry* is one of Poe's most perfectly structured stories. In the first half, he describes the clock-dominated Dutch borough of Vondervotteimittiss. Since the bourgeois inhabitants are interested only in time and sauerkraut, clocks and cabbages dominate their existence. The town itself is evidently built on the regular and well-ordered design of a clock:

> Round the skirts of the valley (which is quite level...), extends a continuous row of sixty little houses. These...look...to the centre of the plain, which is just sixty yards from the front door of each dwelling. Every house has a small garden before it, with a circular path, a sundial, and twenty-four cabbages.... for, time out of mind, the carvers of Vondervotteimittiss have never been able to carve more than two objects—a timepiece and a cabbage.

Each wife carries a Dutch watch in her left hand; the cats and the pigs have repeater watches tied to their tails; and the boys, like their fathers, solemnly smoke a pipe and carry a watch. But each father has the special job of sitting with a grave face and keeping one eye on the great clock that stands in the center of the plain. The great clock is so perfect that the belfry man's job is a sinecure. Anyone who questioned the clock with seven faces would be "considered heretical": "Never was such a place for keeping the true time. When the large clapper thought proper to say 'Twelve o'clock!' all its obedient followers opened their throats simultaneously, and responded like a very echo. In short, the good burghers were fond of their sauerkraut, but then they were proud of their clocks."

At the center of the story, Poe introduces his self-conscious peripeteia or turning point, "I have thus far painted the happy estate of Vondervotteimittiss: alas, that so fair a picture should ever experience a reverse!" It is five minutes before the daily quasi-religious ceremony at the noon hour. (Though Poe's implications are much wider, Gustav Doré's portrayal, in the end papers, of burghers checking their clocks by the noon cannon indicates the type of scene that is being caricatured.) At this very moment, order is shattered in Vondervotteimittiss. The devil appears in the form of a fiddler in black who has not "the remotest idea in the world of such a thing as *keeping time*...." Though the carefree devil is attacking the belfry man up in the "belfry of the House of the Town Council," the burghers are enslaved by their own sense of order, and they follow through with the noonday ceremony:

> "One!" said the clock.
> "Von!" echoed every little old gentleman in every leather-bottomed arm-chair in Vondervotteimittiss. "Von!" said his watch also; "von!" said the watch of his vrow; and "von!" said the watches of the boys, and the little gilt repeaters on the tails of the cat and the pig.

The whole ceremony proceeded according to this well-regulated pattern until the bell said "Twelve": "'Dvelf!' they replied, perfectly satisfied, and dropping their voices."

But at this moment the clock struck "Thirteen": "'*Der Teufel*!' gasped the little old gentlemen, turning pale... it seemed as if old Nick himself had taken possession of every thing in the shape of a time piece. The clocks carved upon the furniture took to dancing as if bewitched, while those upon the mantel-pieces... kept such a continual striking of thirteen, and such a frisking and wriggling of their pendulums as was really horrible to see." The happy little devil figure had brought "din and confusion" into the very temple of order and regularity. Poe's persona concludes the story, "I left the place in disgust, and now appeal for aid to all lovers of correct time and fine kraut. Let us proceed in a body to the borough, and restore the ancient order of things in Vondervotteimittiss by ejecting that little fellow from the steeple."

It might appear that the "little fellow" is the "devil in the horloge" that, in the early days of timekeeping, upset clocks much like the printer's devil still plagues typesetters. But a great deal had changed since the horological revolution; surely, the real devil that Poe questions is the clock itself, symbol of that order and precision through which our bourgeois society has been so thoroughly enslaved.

Tieck mocks the time-oriented nature of the bourgeoisie by allowing two travellers and a sexton to undercut themselves in the amusing "Prologue" to *Kaiser Octavianus*. One of the speeches by the first traveller is typical; his sophism speaks for itself:

> Das ist gewiss, nichts in der ganzen Welt
> Geht über eine recht honette Uhr.
> Warum? Man weiss dann stets in jeder Stunde,
> Wei viel die Glocke eigentlich geschlagen.
> Man ist dann nicht zu spät und nicht zu früh,

When the poet welcomes the allegorical figure of Romance riding on a horse, the three take their leave with speeches through which Tieck further undercuts their clock-oriented values.[22] Tieck also mocks the bourgeoisie amusingly in *Der Gestiefelte Kater*.

Poe's burghers from *Devil in the Belfry* are not an isolated phenomenon, but are to be found elsewhere in his works. The watch, the pipe, and the attention to order and method are their typical hallmarks. In *Man of the Crowd*, "the upper clerks of staunch firms.... wore watches with short gold chains of a substantial and ancient pattern"; in *Adventure of Hans Pfaall*, each burgher "to a man replaced his pipe carefully in the corner of his mouth... puffed, paused, waddled about, and grunted significantly"; and in *The Business Man*, the persona praises, above all, his own "positive appetite for system and regularity." He tells us elsewhere, "in my general habits of accuracy and punctuality, I am not to be beat by a clock."

Despite the comic arabesque and some intentionally misleading allusions, it is clear that the clock is also an ominous symbol in *A Predicament*. Signora Psyche Zenobia—whose name presumably means "given life by Zeus," himself Dieu or Deus the god of day—climbs high into the steeple of "a Gothic cathedral... which towered into the sky." Having "gained the chamber of the belfry," she thrust her head through "an opening in the dial-plate of a gigantic clock [which] must have appeared, from the street, as a large key-hole, such as we see in the face of the French watches." Viewing the clock from this unusual perspective, she notes "the immense size of these hands, the largest of which could not have been less than ten feet in length.... They were of solid steel apparently, and their edges appeared to be sharp."

From where her head rests against the hour hand, she is attracted by the numerals on the dial-plate, and above all the figure *V*: "Turning my head gently to one side, I perceived, to my extreme horror, that the huge, glittering, scimitar-like minute-hand of the clock had, in the course of its hourly revolution, *descended upon my neck*.... Meantime the ponderous and terrific *Scythe of Time* (for I now discovered the literal import of that classical phrase) had not stopped, nor was it likely to stop.... It had already buried its sharp edge a full inch in my flesh." But, much like ourselves, Signora Zenobia is fascinated by the technology that is about to destroy her: "The ticking of the machinery amused me. *Amused me*, I say for my sensations now bordered upon perfect happiness.... The eternal *click-clack*, *click-clack*, *click-clack* of the clock was the most melodious music in my ears."

When the two hands come together—at twenty-five minutes past five (and presumably at the eclipse of Zeus, god of day)—Signora Psyche Zenobia loses her head. However, the body and the mind (or psyche) are still able to communicate, and what they say is humorously reminiscent of the dualism inherent in Cartesian mechanistic philosophy.

In the Kafkaesque *Pit and the Pendulum*, Poe has dropped his dark humor. The mood is closer to that of *The Cask of Amontillado*, but the antagonist is even more sinister because we suffer with the persona. Poe's persona is tied down in a dark prison beside a rat infested "*pit*, typical of hell." Over his head "was the painted figure of Time as he is commonly represented, save that, in lieu of a scythe, he held what, at a casual glance, I supposed to be the pictured image of a huge pendulum, such as we see on antique clocks." This mechanized and much more terrifying version of the sword of Damocles is tipped by a crescent of glittering steel. The razor-sharp edge of the pendulum drops slowly but ineluctably down upon the helpless man, "and the whole *hissed* as it swung through the air." Poe's story foreshadows the terrifying and similarly methodical engine of Kafka's *Penal Colony*. It is a fine piece of symbolism to give part of the clockwork itself not merely the hiss of the snake but also (as in *Predicament*) the cutting edge that acts as the scythe of Time.

In general, it is not in the nature of poets to argue a closely reasoned case. One has to read into the tone, the imagery, and the symbolism of their works. However, in *The Colloquy of Monos and Una*, Poe uses the type of indirection that occurs in Lucian's *Dialogues of the Dead* in order to condemn technological progress through the words of the dead Monos: "One word first, my Una, in regard to man's general condition at this epoch. You will remember that one or two of the wise among our forefathers—wise in fact, although not in the world's esteem—had ventured to doubt the propriety of the term 'improvement,' as applied to the progess of our civilization." Monos can now see that, to survive, humanity should have submitted "to the guidance of the natural laws, rather than attempt their control." It is very difficult to assess the extent to which Romantic poets really understood the dangers of technology in the same sense that the world is now beginning to question its effect on population and pollution. Yet their prophesy of the diabolism inherent in clockwork values is clear enough, and may prove more true than we suspect.

Samuel Butler: The Dangers of Western Technology and Its Concomitant the Well-Regulated Clockwork State

One of the clearest statements on the dangers of what the watch represents comes not from a poet but from the writer of a work that—like *Gulliver's Travels* and *Brave New World*—is an anti-Utopia. Samuel Butler's *Erewhon* (which like Utopia means "nowhere") concerns a land that has learned to limit the use of technology. One hundred years after Sterne's *Tristram*, Butler's attack on the limitations of a mechanically oriented society is far more explicit and perhaps correspondingly less subtle. Butler was particularly concerned with the relationship between mechanism and life. Even before *Erewhon*, he grappled with this both in *Darwin among the Machines* (1863) and his early essay *The Mechanical Creation*.

In the context that "man is committed hopelessly to the machines," Butler's *Mechanical Creation* considered "the probable fate of mankind, if mechanical life should prove ultimately higher than animal." Even the early Butler felt that machines "will breed, and beyond a doubt, varieties and subvarieties of the human race will be developed with a special view to the requirements of certain classes of machinery; we can see the germs of this already in the different aspects of men who attend on different classes of machinery, but they will, as far as we can see, find us always in so many respects serviceable that it would hardly better suit their turn to exterminate us than it would ours to do the like by them"[23] The fitting of men to machines receives further attention in Huxley's *Brave New World*. There, advances in biology permit humans to be bottle bred and adapted through "Neo-Pavolvian conditioning" to the requirements of different classes. No less than "ninety-six identical twins working ninety-six identical machines" can be produced from "A single bokanovskified egg."[24]

Butler's *Erewhon* enlarges considerably on his views regarding the breeding of machines and man's relationship with them. The plot contrivance is that the narrator, Higgs, arrives in a remote Utopia, probably somewhere in New Zealand. He discovers there a society that had turned its back on technology for about four hundred years. It had previously been far more advanced than the European civilisation of the middle of the nineteenth century. Watches, in particular, are banned, and Higgs risks serious recriminations for carrying one. The Erewhonians have something in common with the Brobdingnagians. Neither country has developed gunpowder, and both have discovered clockwork. But a century and a half after *Gulliver*, the potential dangers of clockwork and its related technologies are much more evident. The Erewhonians permit only museums to own such potentially dangerous artifacts as watches and parts of steam engines.

The scene in which Higgs' watch is discovered may remind the reader of the Lilliputians having considered Gulliver's timepiece to be either an animal or "the God that he worships." Butler cannot resist saying "I had thought of Paley, and how he tells us that a savage on seeing a watch would at once conclude that it was designed." Higgs feels, at first, that the "look of horror and dismay on the face of the magistrate" might indicate that the watch was rather "the designer of himself and of the universe." However, he soon realizes that the "expression on the magistrate's face was not of fear, but hatred."[25]

The "Book of the Machines" is an essential part of *Erewhon*, and the bible of those who decided to turn away from the accelerating dangers of technology. Its call is prophetic: "I fear none of the existing machines; what I fear is the extraordinary rapidity with which they are becoming something very different to what they are at present." Butler demonstrates remarkably well the survival of the fittest among machines; their development towards greater and greater self-sufficiency; and the dependency of man on machine that makes him act as their agent for reproduction and evolution. (Man's role is visualized as being comparable to that of an insect when fertilizing plants.) The "Book of the Machines" recognized, in 1872, that the point of no return had already been passed with reference to man's dependence on technology and its related division of labor: "If all machines were to be annihilated at one moment . . . and all machine-made food destroyed so that the race of man should be left naked as it were upon a desert island, we should become extinct within six weeks." Like Swift—with reference to the false religious traditions embroidered into Martin's coat in *Tale of a Tub*—"The Book of the Machines" can only recommend moderation in reversing the flood of technology.

Conclusion

Utopias and anti-Utopias reflect perhaps as well as any works the ideals and fears of their age. In More's *Utopia* (1516) and Andreae's *Christianopolis* (1619), just as in Plato's *Republic*, the ideal is a well-ordered system of government. Fears of the newly emerging capitalist society drive More to suggest an idealized commonwealth and Andreae to project a Christian state. The earlier works are not yet adversely concerned with science and technology, because (in its modern form) it has not sufficiently impinged upon life. Socrates employs the advantage of dividing work into trades as an important analogy throughout *The Republic*. But after the scientific and horological revolution the role of science and technology is fundamental to any consideration of man's estate.

Swift's Gulliver still considers the ordered pastoral condition of the Houyhnhnms as his ideal, but the attack on science, particularly in "Laputa," reflects a new emphasis of satirists that becomes evident with Butler's *Hudibras* (1663) and Shadwell's *Virtuoso* (1676). *Gulliver* (1726) was written at the height of the horological revolution, and this study will have demonstrated why a direct attack on clocks (or clockmakers) could hardly be mounted at that time.

The case with *Erewhon* is different. In Butler, by implication, and more clearly in *Brave New World* and *Nineteen Eighty-Four*, the well-regulated clockwork state is no longer considered an ideal. Only in recent years have the broader masses begun to understand the danger inherent in mankind's overwhelming desire to follow the lead of Western technology. Not only might machines duplicate themselves beyond the capacity of the earth's energy supply, but so might men. Between the time of Christ and the beginning of the scientific and technological revolution, it took sixteen hundred and fifty years for the human population to double to five hundred million. The figure had reached a billion by the time of Malthus; had boubled again to two billion by 1900 A.D.; and is already touching four billion. The world population is predicted to reach seven billion by 2000 A.D., and fifteen billion by 2050 A.D., with some cities containing up to one billion people. No rational being could deny that a continuation of such exponential growth makes Armageddon possible during the lifetime of our children, and without the need for supernatural intervention.

But in the course of the past three centuries Western man has become closely integrated into the clockwork nature of his technological system; he would seem ill suited for reversing, through rational planning, his present conception of progress. It is true that the new technology, in conjunction with the new time-oriented bourgeois society, has been directly responsible for the ever-widening franchise among men of all colors, classes, and creeds. Increasing production demands an increasing market; however, the price for being included in our Western paradise is that all men must eventually accept the same clock-controlled life-style with its related materialistic ethic. Our tragedy would seem to be that, after breaking out of the enclosed medieval world, we have allowed our desire for material progress to transform us into time-oriented automata.

Yet we are barely aware of the extent to which our natures have changed. A society is perhaps best reflected in the metaphors that it uses. The clock analogy became a part of Western man's philosophic systems (indeed, a part of his very way of thinking) as a concomitant of the circumstances leading up to and creating the horological revolution. Like all metaphors that live long enough, it eventually faded into the language. The clock has hands, a face, and much else that derived originally from the organic metaphor. However (in the London vernacular at least), only a man can receive a punch in the clock that may change the look on his dial and even stop his ticker. Our fate frequently depends on which way the pendulum swings, and whether we work in low gear or high gear is often a question of how our train of thoughts makes us tick. The professional man whose speciality is rusty can polish up the subject and be surprised how much more smoothly his work will run. Instead of responding to the good old-fashioned Elizabethan humors, nowadays we are more likely to feel wound up or run down. However, the wheels of life may be running down for more than the individual. Before winding up a company, if top management does not itself have a screw loose, it will check some of the cogs in the organization, and, with the minimum of adjustment, may soon have it running like clockwork. In other areas we might feel the need to oil the wheels of a religious movement, look into the mechanism of a crime syndicate, examine the mainspring of a society's motivation, or regulate a political machine.

At the dawn of modern science, poets, theologians, scientists, and philosophers were rethinking the nature of themselves and their universe in mechanical terms. They needed to visualize those things that cannot be seen through an analogy with something that can. The watch was the color television or computer of that time, not only the pride of author and reader but a point of concrete reference that both could share. As an analogy it was employed as a maid of all work to "explain" many things. But above all its purpose was to serve man by delving into the "burthen of the mystery." A whole spectrum of thinkers used the watch analogy to "explain" the workings of animals, man himself, the state, the mind, the world, the universe, and even God. Moreover, during the horological revolution, the watch became the most important symbol of order—the mechanistic philosophy, the simplification of prose, the regularity of poetry, and the watchmaker God attest to its influence in many spheres.

After about 1760, however, there was a realignment of values. Influenced partly by the developing biological sciences, philosophers opposed the Christian teleological argument, and poets opposed the Augustan clockwork tintinnabulum of rhyme. Apart from the time-oriented bourgeoisie, only theologians continued to accept the order and regularity associated with a watchmaker God. Though Romantic poets tended to be as God-oriented as theologians, they frequently related clockwork order and regularity with the devil. But the Romantics also saw diabolism in clockwork for another reason. It had been the original symbol of the new technological progress, and they were the prophets whose "pastoral" values opposed such progress.

Samuel Butler is probably right when he warns that we probably cannot turn the clock back. Yet if the clock has become the devil, what are we to put in its place? This book does not have the answer. Its more modest purpose has been to document some of the influence of horology on the minds and the literature of men during two crucial centuries.

The clock, in so many ways, is the key machine of the modern industrial age. Much of what we now are derives from our time-oriented technology, and the related clockwork metaphor that dominated our thought processes for so long. The clock and the clock analogy can tell us a great deal about the birth of modern science, and perhaps something too about what makes Western Man tick.

NOTES

For works frequently cited, all references after the first are normally made in the text.

Preface

1. The lack of attention by critics is true even of the period after 1760. Francis D. Klingender's valuable study, *Art and the Industrial Revolution* (London: Paladin, 1972), finds no need either to mention or to portray clocks and watches. In a comparable manner, Herbert L. Sussman's *Victorians and the Machine: The Literary Response to Technology* avoids horology and thinks essentially in terms of "the smoke of the steam engine and the roar of the loom": "machine technology did not truly engage the literary imagination until the coming of the railway" (Cambridge, Mass.: Harvard University Press, 1968), pp. 41 and 9. Perhaps more suprisingly Ivanka Kovacevitch's "The Mechanical Muse: The Impact of Technical Inventions on Eighteenth-century Neoclassical Poetry" (*Huntington Library Quarterly* [1964–65]) also makes no mention of horology.

The Horological Revolution (1660–1760)

1. F. A. B. Ward, *Time Measurement: Historical Review* (London: Science Museum, 1970), p. 15. See also J. D. North, "Monasticism and the First Mechanical Clocks," *The study of Time II*, ed. J. T. Fraser and N. Lawrence (New York: Springer-Verlag, 1975), pp. 381–98.

2. Samuel Johnson, *A Dictionary of the English Language* (London: J. and P. Knapton, 1755), "Horology."

3. Dante Alighieri, *The Divine Comedy*, Carlyle-Okey-Wickstead translation (New York: Modern Library, 1959), p. 466. For further reference to this simile in the closing lines of canto 10, see Ricardo J. Quinones, *The Renaissance Discovery of Time* (Cambridge, Mass.: Harvard University Press, 1972), pp. 37 and 511.

4. Joseph Needham, Wang Ling, and Derek J. de Solla Price, *Heavenly Clockwork: The Great Astronomical Clocks of Medieval China* (Cambridge: University Press, 1960), p. 2, *passim*; and Derek J. de Solla Price, "On the Origin of Clockwork, Perpetual Motion Devices, and the Compass," *United States National Museum Bulletin 218* (Washington, D.C.: Smithsonian Institution, 1959), pp. 82–112.

5. Silvio A. Bedini, "Oriental Concepts of the Measure of Time—The Role of the Mechanical Clock in China and Japan," *The Study of Time II*, ed. Fraser and Lawrence, pp. 451–84.

6. C. B. Drover, "Sand-Glass 'Sand,'" *Antiquarian Horology* 3 (June 1960): 62–67.

7. Cited from Drover, "Sand-Glass," p. 62.

8. D. W. Waters, "Time, Ships and Civilization," *Antiquarian Horology* 4 (June 1963): 82.

9. Cited from Drover, "Sand-Glass," p. 63.

10. Price, "Origin," pp. 91–92, and "Clockwork before the Clock and Timekeepers before Timekeeping," *The Study of Time II*, ed. Fraser and Lawrence.

11. Henry Michel, "Some New Documents in the History of Horology," *Antiquarian Horology* 3 (March 1962): 288–91; C. B. Drover, "The Brussels Miniature," *Antiquarian Horology* 3 (September 1962): 357–61; and Ward, *Time: History*, p. 23.

12. Kenneth Ullyet, *In Quest of Clocks* (London: Hamlyn, 1968), p. 251.

13. F. A. B. Ward, "The Earliest Illustration of a Watch," *Antiquarian Horology* 4 (December 1964): 278 and front cover.

14. Ward, *Time: History*, pp. 24–25.

15. Charles Singer et al., eds., *A History of Technology* (Oxford: Clarendon Press, 1957), 3: 657; Michel, "Some New Documents," pp. 288–91; and Drover, "Brussels," pp. 357–61.

16. Ward, *Time: History*, p. 24; E. J. Tyler reports on "the earliest known clock to show hours, minutes and seconds" of c. 1540 (*Horological Journal* 115 [December 1972]: 19).

17. For an account of automata in the Strasbourg clock see F. C. Haber, "The Cathedral Clock and the Cosmological Clock Metaphor," *The Study of Time II*, ed. Fraser and Lawrence; for some interesting anecdotes about jacks and other automata see Edward J. Wood, *Curiosities of Clocks and Watches* (1866; rpt. Wakefield: EP Publishing Ltd., 1973), pp. 104–107, 20–21, and 132–33.

18. Christiaan Huygens of Zulichem, *Horologium* (1658), trans. Ernest L. Edwardes, in *Antiquarian Horology* 7 (December 1970): 43–44.

19. Waters, p. 85.

20. Eric Bruton, *Clocks and Watches* (Feltham: Hamlyn, 1968), p. 84.

21. Isaac Newton, *Correspondence*, ed. H. W. Turnbull (Cambridge: University Press, 1959), 1: 11.

22. Thomas Sprat, *The History of the Royal Society*, ed. Jackson I. Cope and Harold Whitmore Jones (1667; rpt. St. Louis: Washington University Studies, 1959), p. 382.

23. Huygens, trans. Edwardes, p. 44.

24. Regarding Hooke's rival claim, see A. R. Hall, "Robert Hooke and Horology," *Notes and Records* (London: The Royal Society, 1950), 7: 166–77.

25. Samuel Guye and Henry Michel, *Time & Space*, trans. Diana Dolan and Samuel W. Mitchell (London: Pall Mall Press, 1970), pp. 65–67, and plates 38–44.

26. Francis Wadsworth, "A History of Repeating Watches," *Antiquarian Horology* 4 (September 1965): 364–67; and 5 (December 1965, March 1966, and June 1966): 24–26, 48–52, and 90–92. I am indebted to Wadsworth for much of the previous information on repeating watches.

27. Singer, ed. *History of Technology*, 3: 652.

28. For valuable insights into the technical developments related to time measurement, see in particular "The Empirical Search" in Fraser's *Of Time, Passion, and Knowledge* (New York: George Braziller, 1975). *The Study of Time II*, ed. Fraser and Lawrence, has a "Special Session on Timekeepers and Time," to which Fraser contributes "Clockmaking—The Most General Trade"; H. Alan Lloyd provides a brief but useful historical outline in "Timekeepers—An Historical Sketch," *The Voices of Time*, ed. J. T. Fraser (New York: George Braziller, 1966).

The Horological Revolution and the Industrial Revolution

1. Michael Hurst, "The First Twelve Years of the Pendulum Clock," *Antiquarian Horology* 6 (June 1969): 146.
2. G. H. Baillie and C. A. Ilbert, *Britten's Old Clocks and Watches and Their Makers*, 7th ed. (New York: Bonanza Books, 1956), p. 66; Ward, *Time: History*, pp. 17 and 1.
3. Hurst, pp. 151–52.
4. Adam Smith, *An Inquiry into the Nature and Causes of the Wealth of Nations*, ed. Edwin Cannan (New York: Modern Library, 1937), p. 243.
5. Ibid., pp. 128 and 123.
6. Lawrence Wright, *Clockwork Man* (London: Elek Books, 1968), p. 87.
7. R. W. Symonds, *Thomas Tompion: His Life and Work* (London: Spring Books, 1969), p. 237.
8. Ibid., pp. 237–38.
9. Charles Babbage, *On the Economy of Machinery and Manufactures* (London: Charles Knight, 1832), p. 233.
10. Frank West, "A Short History of the British Horological Institute," *Horological Journal* 113 (May 1971): 3.
11. See in particular Edwin A. Battison's contention that the disruption of foreign trade during the first two decades of the nineteenth century "provided a great spur to domestic industry" (*The American Clock 1725–1865* [Greenwich, Conn.: N.Y. Graphic Society, 1973], p. 17).
12. Symonds, p. 232.
13. Richard Foster Jones, *Ancients and Moderns*, 2nd ed. (Berkeley and Los Angeles: University of California Press, 1961), pp. 84, 87–88, 114–18, *passim.*
14. Stephen J. Gendzier, trans., "Preliminary Discourse," in *Denis Diderot's The Encyclopedia: Selections* (New York: Harper and Row, 1967), p. 41.
15. Denis Diderot and Jean le Rond d'Alembert, eds. *Encyclopédie* (Neufchastel [acutually Paris]: Samuel Faulche &c Cie, 1751–65), 8: 298–310.
16. Samuel L. Macey, "Work Study before Taylor: An Examination of Certain Preconditions for Time and Motion Study That Began in the Seventeenth Century," *Work Study and Management Services* 18 (October 1974): 530–36.
17. Ferchault de Réamur, *Art de l'Épinglier . . . Avec des Additions de M. Duhamel du Monceau, & des Remarques extraites des Mémoires de M. Perronet* [Paris, 1762], pp. 43–44; and Babbage, p. 146.
18. A. R. Hall, "Robert Hooke and Horology," *Notes and Records* (London: The Royal Society, 1950), 7: 167–77; and Symonds, p. 132.
19. Jean-Claude Sabrier and Andre Imbert, "An Early Anonymous Swiss Self-Winding Watch," *Antiquarian Horology* 7 (March, 1972): 526–27.

20. H. von Bertele, "The Origin of the Differential Gear," *Antiquarian Horology* 2 (December 1956): 14–15.

21. Macey, "Work Study before Taylor," *loc. cit.*; R. M. Currie, *Work Study* (London: Pitman, 1968), p. 21.

22. A. Alan Lloyd, "Samuel Watson," *Antiquarian Horology* 1 (December 1954): 60–61.

23. Maurice Daumas, *Scientific Instruments of the Seventeenth and Eighteenth Centuries and their Makers*, trans. Mary Holbrook (London: B. T. Batsford, 1972), pp. 116–17.

24. Singer et al. eds., *History of Technology* (Oxford: Clarendon Press, 1957), 3: 671.

25. Eric Bruton, *Clocks and Watches* (Feltham: Hamlyn, 1968), p. 47.

26. Paul Mantoux, *The Industrial Revolution in the Eighteenth Century* (London: Jonathan Cape, 1961), p. 296.

27. Ibid., pp. 216–17, 228–32.

28. Babbage, pp. 278–79.

29. A. E. Musson and Eric Robinson, *Science and Technology in the Industrial Revolution* (Manchester: Manchester University Press, 1969), pp. 435, 457, 436–37, 24, 438, and 143; Josiah Wedgwood, *Selected Letters*, ed. Ann Finer and George Savage (London: Cory, Adams & Mackay, 1965), plate 10; and J. G. Crowther, *Scientists of the Industrial Revolution* (London: Cresset Press, 1962), p. 130.

The Horological Revolution and Society

1. G. H. Baillie, *Clocks and Watches: An Historical Bibliography* (London: N.A.G. Press, 1951), p. 103. Baillie questions one of the horological designs of Leibniz.

2. W. Marshall, "Voltaire: Politician, Statesman, Financier, Author," *Horological Journal* 113 (July 1971), 19–20.

3. R. W. Symonds, *Thomas Tompion: His Life and Work* (London: Spring Books, 1969), p. 232.

4. Samuel Pepys, *The Diary*, ed. Henry B. Wheatley (London: G. Bell and Sons, 1924), 4 Sept.1663; 6 Sept. 1667; 22 Sept. 1666; 24 July 1661; 14 July 1665; and 12 May 1665. John Evelyn, *Diary*, ed. William Bray (London: Bickers and Son, 1879), 2: 119 and 133; 1: 127, 138, 171, and 238; and 2: 125.

5. Pepys, 22 Dec. 1665.
6. Thomas Sprat, *The History of the Royal Society*, ed. Jackson I. Cope and Harold Whitmore Jones (1667; rpt. St. Louis: Washington University Studies, 1959), pp. 127, 246–47, 250, 254–55, 317, and 388.
7. Ibid., pp. 388–89.
8. John George Keysler, *Travels*, 2nd ed. (London: A. Linde, 1756), 1: 301, 100, 251 and 280.
9. Boyle Papers, Vol. ii, f. 141^{r} and 141^{v}. See Laurens Laudan, "The Clock Metaphor and Probabilism: The Impact of Descartes on English Methodological Thought, 1650–1665," *Annals of Science* 22 (June 1966): 89.
10. Robert Boyle, *Certain Physiological Essays* (London: Henry Herringman, 1661), p. 17.
11. Robert Boyle, *Works* (London: W. Johnston and others, 1772), 1: 62–63.
12. Sir Kenelm Digby, *Two Treatises* (Paris: Gilles Blaizot, 1644), chapters 36–38. See in particular p. 208.
13. Boyle, *Works*, 3: 260–61.
14. Ibid., 1: 106.
15. Leonora Cohen Rosenfield, *From Beast-Machine to Man-Machine*, enlarged ed. (New York: Octagon Books, 1968), pp. 281–83.
16. Boyle, *Works*, 5: 15; 3: 464; 4: 786; 5: 47.
17. Ibid., 6: 210 and 111.
18. Ibid., 3: 397–99.
19. Ibid., 2: 374–76.
20. A. J. Turner, "Christopher Wren and the Wadham Clock," *Antiquarian Horology* 7 (June 1971): 229–30; Charles K. Aked, "William Derham and 'The Artificial Clockmaker,'" *Antiquarian Horology* 6 (March, June, and September 1970): 362–69, 416–27, and 495–505; and John R. Millburn, "Some Horological Extracts from Stukeley's Diaries," *Antiquarian Horology* 6 (September 1969): 206–11.
21. Oliver Goldsmith, *Collected Works*, ed. Arthur Friedman (Oxford: Clarendon Press, 1966), 2: 42–44.
22. Daniel Defoe, *Roxana, The Fortunate Mistress* (New York: Doubleday, 1964), p. 60.
23. Daniel Defoe, *The Fortunes and Misfortunes of the Famous Moll Flanders* (New York: Holt, Rinehart and Winston, 1949), p. 266. The same point is repeated elsewhere; see also pp. 217 and 259.
24. Daniel Defoe, *The History and Remarkable Life of the Truly Honourable Col. Jaque Commonly Call'd Col. Jack*, ed. Samuel Holt Monk (London: Oxford University Press, 1970), pp. 63–65.
25. Frances Parthenope Verney and Margaret M. Verney, *Memoirs of the Verney Family during the Seventeenth Century*, 2nd ed. (London: Longmans, Green and Co., 1904), 2: 135.
26. Samuel L. Macey, "The Time Schemes in *Moll Flanders*," *Notes and Queries* 214 (September 1969): 336–37.

27. Defoe, *Moll*, pp. 72–73, 155, and 206.
28. G. H. Baillie and C. A. Ilbert, *Britten's Old Clocks and Watches and Their Makers*, 7th ed. (New York: Bonanza Books, 1956), p. 39.
29. Symonds, pp. 309–11 and 237.
30. Defoe, *Moll*, pp. 207 and 287.
31. Ibid., pp. 232, 241, 234, 218, 212, 267, 272, and 208.
32. Ibid., pp. 261, 325, 351, and 349.
33. See Bernard Falk, *Thomas Rowlandson: His Life and Art* (New York: The Beechhurst Press, 1952), *passim*; and James Gillray, *Works* (London: Henry G. Bohn, 1851), *passim*. Gillray frequently alludes to opulence and loot, but he uses gold coins (in great profusion), and not watches.
34. Francis D. Klingender, *Art and the Industrial Revolution*, ed. Arthur Elton (London: Paladin, 1972), p. 46.
35. Karl-Adolf Knappe, *Dürer: The Complete Engravings, Etchings and Woodcuts* (New York: Harry N. Abrams, n.d.), p. xliv.
36. Ronald Paulson, *Hogarth: His Life, Art, and Times* (New Haven and London: Yale University Press, 1971), 1: 48 and 513. See also 1: 50, 176, and 514. I am indebted throughout this portion of the chapter both to the *Life* and to Paulson's other definitive work: *Hogarth's Graphic Works*, rev. ed. (New Haven: Yale University Press, 1970).
37. Paulson, *Life*, 2: 244; 1: 494, 433, and 554.
38. Ibid., 1: 291.
39. Ibid., 1: 492; and *Works*, 1: 133.
40. Paulson, *Life*, 2: 267–68.
41. Ibid., 1: 234.
42. Ibid., 2: 175.
43. Klingender, plates 52ff., and pp. 106, 109–10, 117, and 128–29. See also Boswell's *London Journal*, 15 Dec. 1762.
44. M. W. Flinn, *Origins of the Industrial Revolution* ([London]: Longmans, 1966), pp. 71ff.

Philosophers and the Clock Metaphor

1. Francis Bacon, *The Works*, ed. James Spedding et al. (Cambridge: The Riverside Press, 1863), 8: 83.

2. Cited from R. F. Jones, *Ancients and Moderns*, 2nd ed. (Berkeley and Los Angeles: University of California Press, 1961), p.55.

3. Bacon, *Works*, 5: 410.

4. John Webster, *Academiarum Examen* (London: Giles Calvert, 1654), p. 19.

5. Ibid., pp. 105–106. George Hakewell takes a comparable view: "It is . . . a . . . worthy endeavour of my Lord of *S. Albanes* so to mixe and temper practice & speculation together, that they may march hand in hand. . . . *Speculation* by precepts and infallible conclusions preparing a way to *Practice*, and *Practice* againe perfecting *Speculation*" (*An Apologie of the Power and Providence of God in the Government of the World* [Oxford: John Lichfield and William Turner, 1627] p. 246).

6. See Samuel L. Macey, "On Dividing the Loot: The Delegation of Power," *The Yale Review* 61 (Spring 1972): 396–406.

7. Bacon, *Works*, 6: 96; and 2: 104–105.

8. Ibid., 8: 169. Bacon says elsewhere: "Among the parts of history which I have mentioned, the history of Arts is of most use. . . . Upon this history therefore, mechanical and illiberal as it may seem (all fineness and daintiness set aside) the greatest diligence must be bestowed" (8: 363); and "most writers of natural history think it enough to make a history of animals or plants or minerals, without mentioning the experiments of mechanical arts (which are far the most important for philosophy) . . ." (10: 407).

9. Bacon, *Works*, 13: 129–31.

10. Thomas Sprat, *History of the Royal Society*, ed. Jackson I. Cope and Harold Whitmore Jones (St. Louis: Washington University Studies, 1959) pp. 129, 118–19, *passim*.

11. Ibid., p. 329.

12. Robert Boyle, *Works* (London: W. Johnston and others, 1772), 3: 397.

13. Ibid., 3: 397–99.

14. I have been concerned elsewhere with the way in which this ever-broadening franchise is reflected in literature (Samuel L. Macey, "Nonheroic Tragedy: A Pedigree for American Tragic Drama," *Comparative Literature Studies* 6 [Winter 1969]: 1–19).

15. I am much indebted here as elsewhere to *Ancients and Moderns*, the pioneer work of R. F. Jones.

16. Bernard Le Boivier de Fontenelle, *A Plurality of Worlds*, trans. John Glanvill (1688; rpt. London: Nonsuch Press, 1929), p. 17.

17. Boyle, *Works*, 2: 39–40.

18. Ralph Cudworth, *A Treatise Concerning Eternal and Immutable Morality* (London: James and John Knapton, 1731), pp. 154–55 and 162.

19. J. D. McFarland, *Kant's Concept of Teleology* (Edinburgh: University of Edinburgh Press, 1970), pp. 45–49.

20. Edward, Lord Herbert of Cherbury, *The Ancient Religion of the*

Gentiles (London: John Nutt, 1705), p. 257.

21. Nieuwentyt, *The Religious Philosopher*, trans. John Chamberlayne (London: J. Senex and W. Taylor, 1718), p. xlvi. See also p. lxviii.

22. Bernard Mandeville, *The Fable of the Bees*, ed. F. B. Kaye (Oxford: Clarendon Press, 1924), 2: 310.

23. John Ray, *The Wisdom of God Manifested in the Works of the Creation*, 9th ed. (London: William and John Innys, 1727), p. 35.

24. Nicole Oresme, *Le Livre du ciel et du monde*, ed. Albert D. Menut and Alexander J. Denomy, trans. Albert D. Menut (Madison: University of Wisconsin Press, 1968), p. 289.

25. H. G. Alexander, ed. *The Leibniz-Clarke Correspondence* (Manchester: Manchester University Press, 1956), pp. 11–12.

26. [John Wilkins], *A Discourse concerning a New Planet* (London: John Maynard, 1640), p. 204.

27. Henry Power, *Experimental Philosophy* (London: John Martin and James Allestry, 1663), preface, pp. 7–10, and 17.

28. Boyle, *Works*, 4: 49.

29. Bolingbroke, *Works* (Philadelphia: Carey and Hart, 1841), 3: 281.

30. René Descartes, *Discourse on Method*, trans. F. E. Sutcliffe (Harmondsworth: Penguin, 1968), pp. 73–74.

31. Ibid., pp. 75–76.

32. René Descartes, *Philosophical Letters*, trans. Anthony Kenny (Oxford: Clarendon Press, 1970), pp. 53–54.

33. Ibid., p. 244.

34. Ibid., pp. 207–208.

35. For the history of animal mechanism in France see: Leonora Cohen Rosenfield, *From Beast-Machine to Man-Machine*, enlarged ed. (New York: Octagon Books, 1968); Robert Lenoble, *Mersenne; ou, La naissance du mécanisme* (Paris: J. Vrin, 1943); Heikki Kirkinen, *Les Origines de la conception moderne de l'homme-machine* (Helsinki, 1960); and Aram Vartanian [ed.], *La Mettrie's L'Homme-Machine* (critical edition with an introductory monograph) (Princeton: Princeton University Press, 1960).

36. Lenoble, *Mersenne*, pp. 318–19.

37. Rosenfield, *Beast-Machine*, p. 21.

38. René Descartes, *Oeuvres*, issued by Charles Adam and Paul Tannery (Paris: Leopold Cerf, 1904), 11: 202.

39. Johann Clauberg, *Opera Omnia Philosophica* (1691; facsimile rpt. Hildesheim: Georg Olms, 1968), 1: 184; and Descartes, *Oeuvres* (Adam and Tannery), 11: 330–31. For Donne and Herbert, see chapter 6, notes 43 and 44.

40. Cited from Rosenfield, *Beast Machine*, pp. 241–42.

41. Power, *Experimental Philosophy*, preface, pp. 7–10.

42. Robert Hooke, *The Posthumous Works*, introduction by Richard S. Westfall (1705; facsimile rpt. New York and London: Johnson Reprint Corporation, 1969), p. 39.

43. Boyle, *Works*, 5: 215.

44. Cited from Rosenfield, *Beast-Machine*, pp. 267, and 265–66.

45. René Descartes, *The Philosophical Works*, trans. Elizabeth S. Haldane, and G. R. T. Ross (Cambridge: University Press, 1967), 1: 299–300.

46. Sir Kenelm Digby, *Two Treatises* (Paris: Gilles Blaizot, 1644), p. 208. The material referred to has been taken from chapters 36–38.

47. René Descartes, *L'Homme* and *Un Traité de la formation du foetus* (Paris: Charles Angot, 1644), p. 112.

48. Ibid., pp. 106–107. Though Hobbes was influenced by Descartes, the clock image comes much more readily to the latter, who clearly thought throughout his life in mechanical terms. Schrynemakers, for example, points to the possibility of Descartes being involved in the important substitution of the chain for the cord in weight-driven clocks that took place around 1640 ("Descartes and the Weight-Driven Chain-Clock," *Isis* 60 [Summer 1969]: 233–36). Guye and Michel credit Gruet with the comparable change from catgut to chain in the fusee after 1650 (*Time & Space*, pp. 77–78).

49. Nicolas Malebranche, *De la Recherche de la vérité*, ed. Genevieve Lewis (Paris: J. Vrin, 1945), 2: 255.

50. Thomas Hobbes, *Leviathan*, introduction by A. D. Lindsay (1651; London: J. M. Dent, 1965), pp. 367–68.

51. J. A. Passmore, *Ralph Cudworth* (Cambridge: University Press, 1951), p. 3.

52. Ralph Cudworth, *The True Intellectual System of the Universe* (London: Richard Royston, 1678), pp. 846–47.

53. Boyle, *Works*, 6: 511.

54. Sprat, *History*, pp. 82–83.

55. John Locke, *Works*, ed. J. A. St. John (1877; rpt. New York: Books for Libraries Press, 1969), 1: 118. See also John Locke, *Some Thoughts Concerning Education* (1693; rpt. Scolar Press: Menston, 1970), p. 158.

56. Leibniz, *The Monadology*, trans. and ed. Robert Latta (London: Oxford University Press, 1898), pp. 331–36.

57. Benedict de Spinoza, *The Chief Works*, trans. R. H. M. Elwes (New York: Dover Publications, 1951), 2: 32. Spinoza's "automa spirituale" is translated by Elwes as "immaterial automaton," and in Andrew Boyle's translation (Everyman Library) as "spiritual automaton."

58. La Mettrie, *Histoire naturelle de l'âme* (La Haye, 1745), pp. 150–51, and 249–50; and *Oeuvres philosophiques* (Londres [Berlin], 1751) *Abrégé des Systêmes*, pp. 237–38. Cited from Aram Vartanian to whom I am indebted for noting La Mettrie's changes (*La Mettrie's L'Homme Machine*, ed. Vartanian, pp. 47–48).

59. David Hartley, *Observations on Man* (London: James Leake and William Frederick, 1749), part 1, pp. 5–6.

60. Ibid., part 1, pp. iii [3], 413, 500, and 504.
61. *La Mettrie's L'Homme Machine*, ed. Vartanian, pp. 149, 186, and 195.
62. Ibid., p. 190.
63. Ibid., loc. cit.
64. Descartes, *Works*, Adam and Tannery, 9: 322; Descartes, *Works*, Haldane and Ross, 1: 289. Laurens Laudan has referred to some of the examples of this analogy in Descartes, Boyle, Glanvill, Power, and Locke ("The Nature and Sources of Locke's Views on Hypotheses," *Journal of the History of Ideas* [1967], pp. 220–22); and Ian Donaldson has compared the analogy in Locke, Johnson, and Richardson ("The Clockwork Novel: Three Notes on an Eighteenth-Century Analogy," *Review of English Studies* 21 [1970]: 14–22).
65. Bougeant, *A Philosophical Amusement upon the Language of Beasts* (London: T. Cooper, 1739), p. 5.
66. Joseph Glanvill, *Scepsis Scientifica* (London, E. Cotes, 1665), pp. 155–56. An earlier version, *The Vanity of Dogmatizing*, appeared in 1661.
67. Ibid., p. 133.
68. Power, *Experimental Philosophy*, preface, p. 17.
69. Ibid., pp. 192–93.
70. Boyle, *Works*, 5: 443.
71. Ibid., 2: 45.
72. Ibid., 2: 45–46.
73. Boyle Papers, Vol. ii, f. 141^{v}.
74. Boyle, *Works*, 2: 71.
75. Ibid., 2: 46.
76. Locke, *Works*, 2: 41–42. Locke's language moves readily into the mechanical metaphor. See, for example, John Locke, *Two Treatises of Government*, ed. Peter Laslett, revised ed. (New York and London: Cambridge University Press, 1963), pp. 214–15, 311, and 347.
77. George Berkeley, *Works*, ed. A. A. Luce and T. E. Jessop (London: Thomas Nelson, 1948–56), 5: 111–12.
78. George Berkeley, *A Treatise concerning the Principles of Human Knowledge* (first printed 1710; London: Jacob Tonson, 1734), p. 62.
79. Ibid., pp. 64 and 60.
80. John Ray, *Wisdom of God*, p. 25.
81. Sir Isaac Newton, *Mathematical Principles*, trans. Andrew Motte, ed. Florian Cajori (trans. 1729; rpt. Berkeley and Los Angeles: University of California Press, 1962), Cotes' "Preface to the Second Edition," pp. xxvii–xxviii.
82. Cited from Donaldson, "Clockwork Novel," p. 14.
83. James Boswell, *Life of Johnson*, ed. George Birkbeck Hill and L. F. Powell (Oxford: Clarendon Press, 1934–50), 2: 174.
84. Ibid., 2: 49. Hume, in the *Dialogues concerning Natural Religion*, deals with similar ideas in terms of the clock metaphor, but since his

purpose would seem to involve a reaction to the analogy it has not been included here.

85. Simon Patrick, *A Brief Account of the New Sect of Latitude-Men*, introduction by T. A. Birrell (Los Angeles: Augustan Reprint Society, 1963), pp. 14–19, and 22. Since he was associated with the Cambridge Platonists, Patrick's remarks on the closing page of the pamphlet are of particular interest: "Christian Religion was never bred up in the *Peripatetick School*... let her old loving Nurse the *Platonick Philosophy* be admitted again to her family; nor is there any cause to doubt but the Mechanick also will be faithful to her...."

86. Hobbes, *Leviathan*, p. 1.

87. Basil Willey, *The Seventeenth Century Background* (New York: Doubleday, 1953), pp. 114–15.

88. Anthony, Earl of Shaftesbury, *Characteristicks* (1714; facsimile rpt. Farnborough: Gregg International Publishers, 1968), 1: 116.

89. Ibid., 1: 115–16. In "An Enquiry concerning Virtue," Shaftesbury uses the standard simile of animals acting like "a simple Mechanism, an Engine, or Piece of Clockwork" (Ibid., 2: 86).

90. Walter Charleton, *Physiologia Epicuro-Gassendo-Charltonia*, introduction by Robert Hugh Kargon (1654; facsimile rpt. New York and London: Johnson Reprint Corporation, 1966), p. 114.

91. John Trapp, *A New Comment upon the Pentateuch* (London: Timothy Garthwait, 1650), p. 109.

92. Anon., *Humane Industry: Or, A History of Most Manual Arts*, p. 8.

93. Boyle, *Works*, 2: 71.

94. Ibid., 5: 136.

95. Boswell, *Life of Johnson*, ed. Hill and Powell, 2: 49.

The Watchmaker God

1. Sir Matthew Hale, *The Primitive Origination of Mankind* (London: William Shrowsbury, 1677), pp. 1ff.

2. John Locke, *Works*, ed. J. A. St. John (1877; rpt. New York: Books for Libraries Press, 1969), 1: 431–32.

3. Gottfried Wilhelm Leibniz, *Philosophical Papers and Letters*, trans. Leroy E. Loemaker, 2nd ed. (Dordrecht: D. Riedel, 1969), p. 648. See also p. 652.
4. Ibid., p. 675.
5. Ibid., p. 110.
6. Sir Isaac Newton, *Optics*, forward by Albert Einstein (1730: 4th ed. New York: Dover Publications, 1952), pp. 397–98, and 402.
7. H. G. Alexander, ed., *The Leibniz-Clarke Correspondence* (Manchester: Manchester University Press, 1956), pp. 11–12. Perpetual motion, apparently unknown to the ancients, now enjoyed a considerable vogue; by way of refutation, the interest gave rise to the law of the conservation of energy.
8. Ibid., pp. 13–14.
9. Ibid., p. 18.
10. Ibid., p. 22.
11. Ibid., pp. 53, and 110.
12. Ibid., pp. 85–86.
13. Leibniz, *The Monadology*, trans. and ed. Robert Latta (London: Oxford University Press, 1898), p. 333; see also pp. 320–21, and 331–36.
14. Sir Leslie Stephen, *History of English Thought in the Eighteenth Century*, 3rd ed. (New York: G. P. Putnam's Sons, 1902), 1: 408–409; and Henry Hallam, *Introduction to the Literature of Europe*, 4th ed. (London: John Murray, 1854), 2: 384–85.
15. R. R. Palmer, *A History of the Modern World*, 2nd ed. (New York: Alfred A. Knopf, 1960), pp. 290–91.
16. Philip of Mornay, *A Woorke concerning the Trewnesse of the Christian Religion*, trans. Sir Philip Sidney and Arthur Golding (London: Thomas Cadman, 1587), pp. 99–100 and 97; John Smith, "Of the Existence and Nature of God," *Select Discourses* (London: W. Morden, 1660), p. 131; Lord Herbert of Cherbury, *The Ancient Religion of the Gentiles* (London: John Nutt, 1705), p. 257; John Spencer, *A Discourse concerning Prodigies* (Cambridge: William Graves, 1663), preface; N. Fairfax, *A Treatise of the Bulk and Selvedge of the World* (London: Robert Boulter, 1674), pp. 75–76.
17. Hale, *Primitive Origination*, pp. 340–42.
18. Boyle Papers, Vol. ii, f. 142[v]; see also Robert Boyle, *Works* (London: W. Johnston and others, 1772), 5: 130 and 2: 40.
19. Cudworth, *Eternal and Immutable Morality*, p. 175.
20. Sir Richard Blackmore, *Creation*, in *Works of the English Poets*, ed. Alexander Chalmers (London: J. Johnson and others, 1810), 10: 337, 353, and 356.
21. W. Derham, *Astro-Theology: Or a Demonstration of the Being and Attributes of God, from a Survey of the Heavens* (London: W. Innys, 1715), pp. 67–68, and 208–11.
22. Edward Search [Abraham Tucker], *The Light of Nature Pursued* (London: T. Payne, 1768), 2: 94–97; also 1: 2–3 and 58–59.

23. Nieuwentyt, *Religious Philosopher*, trans. John Chamberlayne (London: J. Senex and W. Taylor, 1718), preface, p. lxv; and William Paley, *Natural Theology: Or, Evidences of the Existence and Attributes of the Deity, Collected from the Appearances of Nature*, 4th ed. (London: R. Faulder, 1803), pp. 2–3. Wood's lengthy passages from John Martin's *Mechanicus ana Flaven; or The Watch Spiritualized* (1763) show how easily (though pedantically) the watch could lend itself to religious "similitudes" (*Curiosities*, pp. 329–39).

24. F. Le Gros Clark [ed.], *Paley's Natural Theology. Revised to Harmonize with Modern Science* (London: The Christian Evidence Committee of the Society for Promoting Christian Knowledge, 1885).

25. Paley, *Natural Theology*, pp. 3–4.

26. Ibid., p. 9.

27. Ibid., p. 473.

28. David Hume, *Dialogues Concerning Natural Religion*, ed. Norman Kemp Smith, 2nd ed. (New York: Social Sciences Publishers, 1948), p. 143.

29. Ibid., p. 176.

30. Ibid. See in particular pp. 143, 146, 171–72, 176, 180, 190, and 216–17.

31. Boyle, *Works*, 5: 136–37.

32. Locke, *Works*, 1: 460–63.

33. Leibniz, *Philosophical Papers*, p. 456.

34. Aristotle, *De Motu Animalium* 7, trans. A. S. L. Farquharson, in *Works*, ed. J. A. Smith and W. D. Ross (Oxford: Clarendon Press, 1912), 5: 701b.

35. Leibniz, *Philosophical Papers*, p. 637.

36. Ibid., p. 217.

37. Ibid., p. 649.

38. Cudworth, *Eternal and Immutable Morality*, p. 171; see also p. 177.

39. Joshua C. Gregory, "The Animate and Mechanical Models of Reality," *Journal of Philosophical Studies* (1927), p. 302.

40. Ibid., p. 314.

41. Cited in J. D. McFarland, *Kant's Concept of Teleology*, (Edinburgh: University of Edinburgh Press, 1970), p. 107.

42. Ibid., p. 139.

43. Hartley gives credit to Locke (see *Essay Concerning Human Understanding* II. xxxiii), and John Gay, cousin of the famous writer (see [John Gay] "Preliminary Dissertation" prefixed to: William King, *An Essay on the Origin of Evil* [London: W. Thurlbourn, 1731], pp. xxx–xxxiii). The idea is implicit in Hobbes, *Leviathan*, I.iii, and is mentioned briefly in Aristotle, *De Motu Animalium 8*, in *Works*, 5: 701b. See also Plato, *Republic*, book 4.

44. Erasmus Darwin, *Zoonomia* (London: J. Johnson, 1794), p. 480. I am indebted to Dr. Robert M. Young—of the Wellcome Unit for the

History of Medicine, University of Cambridge—for pointing out the relationship between Hartley and Darwin.

45. McFarland, *Kant's Concept of Teleology*, p. 48.

46. Cited from John Fiske's *The Idea of God*, in John Haynes Holmes, "A Struggling God," *My Idea of God*, ed. Joseph Fort Newton (Boston: Little, Brown, and Company, 1926), p. 112.

Poets and the Clock Metaphor

1. Saint Thomas Aquinas, "Psychology of Human Acts," *Summa Theologica.* Latin text and English translation by Thomas Gilby (London: Eyre and Spottiswoode, 1970), 17: 129. The relevant Latin text reads: ". . . idem apparet in motibus horologiorum. . . ."

2. Dante Alighieri, *The Divine Comedy*, Carlyle-Okey-Wickstead translation (New York: Modern Library, 1959), p. 466 (closing lines of canto 10).

3. Nicole Oresme, *Le Livre du ciel et du monde*, ed. Albert D. Menut and Alexander J. Denomy; trans. Albert D. Menut (Madison: University of Wisconsin Press, 1968), p. 289.

4. Froissart, *Oeuvres* (Brussels, Victor Devaux, 1870), 1: 53–86. (Translation mine.) Baillie notes that, as in Dondi's clock (1364), the dial turns once in twenty-four hours while the pointer remains stationary (*Clocks and Watches: An Historical Bibliography* [London: N. A. G. Press, 1951], p. 3).

5. Cited from G. H. Baillie and C. A. Ilbert, *Britten's old Clocks and Watches and Their Makers*, 7th ed. (New York: Bonanza Books, 1956), p. 18.

6. Geoffrey Chaucer, *The Canterbury Tales*, in *Works*, ed. F. N. Robinson, 2nd ed. (Boston: Houghton Mifflin, 1961), p. 199, line 2854. But see also B. H. Bronson, "Concerning Houres Twelve," *Modern Language Notes* 68 (December 1953): 515–21.

7. Baillie, *Clocks; Bibliography*, p. 3

8. Derek J. Price, ed. *The Equatorie of the Planetis* (Cambridge: University Press, 1955), pp. 3–4, *passim.*

9. There would appear to be only one other reference to "houre

inequal," *Knight's Tale*, p. 39, line 2271.

10. Sir Thomas More, *Utopia*, ed. Edward Surtz (New Haven and London: Yale University Press, 1964), p. 70.

11. Christopher Marlowe, *Faustus*, ed. W. W. Greg (Oxford: Clarendon Press, 1950), A text, lines 1456–60.

12. Although my search has not been exhaustive, I have found a remarkable absence of horological references even in a poet as late as Spenser, despite the complex time pattern in his *Epithalamion*. By way of contrast with Defoe and Hogarth, Greene's "Conny-catching" tracts (1592) are essentially concerned with matters other than the taking of watches.

13. Baillie and Ilbert, *Britten's*, p. 43.

14. William Shakespeare, *Works*, ed. G. B. Harrison (New York: Harcourt, Brace, 1958), 5.2.18.

15. Richard Glasser shows that in the French language there is a distinct development towards terms indicating a more precise measurement of time (*Time in French Life and Thought*, trans. C. G. Pearson [Manchester: University Press, 1972], pp. 75–77, *passim*). Research would surely demonstrate a comparable development in the English language.

16. Guil. Hormani, *Vulgaria* (London, 1519), facing p. 232. (Some for a tryfull/pley the devyll in the orlege.") See also John Heywood, *Proverbs and Epigrams* (1562; rpt. n. p.: The Spenser Society, 1867), p. 149; John Heywood, *A Dialogue of Proverbs*, ed. Rudolph Habenicht (Berkeley and Los Angeles: University of California Press, 1963), p. 147; and R. Harvey, *Plaine Percevall* (London, [1590?]), pp. 17–18.

17. Brents Stirling, *The Populace in Shakespeare* (New York: Columbia University Press, 1949), *passim*.

18. William Shakespeare, *The Second Part of the History of Henry IV*, ed. John Dover Wilson (Cambridge: University Press, 1946), p. 169.

19. John Dryden, *Works*, ed. Walter Scott (London: James Ballantyne, 1808), 13: 246. Boswell uses a comparable idiom, "dipped my machine in the Canal and performed most manfully" (*London Journal*, 4 June 1763).

20. John Donne, *The Poems*, ed. Herbert J. C. Grierson (London: Oxford University Press, 1912), 1: 235.

21. Sir John Suckling, *Aglaura*, in *Works*, ed. (Plays) L.A. Beaurline (Oxford: Clarendon Press, 1971), 2: 95.

22. Alexander Pope, *Poems* (Twickenham edition) ed. E. Audra and Aubrey Williams (London: Methuen, 1961), 1: 239–40. All poetry references will be from the Twickenham edition.

23. Samuel Johnson, *Letters*, ed. George Birkbeck Hill (Oxford: Clarendon Press, 1892), 2: 415 (Letter to Francesco Sastres, 21 August 1784).

24. John Preston, *Sermons* (London: Leonard Greene, 1630), pp. 18–19.

25. J. Goodman, *The Penitent Pardoned* (London: R. Royston, 1679), p. 21.
26. Philip Dormer Stanhope, Earl of Chesterfield, *Letters*, ed. John Bradshaw (London: Swan Sonnenschein, 1893), 1: 90.
27. Alexander Pope, *Works*, ed. Whitwell Elwin and William John Courthope (London: John Murray, 1886), 10: 550. This edition will be used for references to the prose works.
28. Robert Southey, *Essays, Moral and Political* (London: John Murray, 1832), 2: 23.
29. Robert Browning, *Works* (New York: Barnes and Noble, 1966), 4: 267.
30. Donne, *Poems*, ed. Grierson, 1: 129.
31. Pope, *Poems* (Twickenham), 6: 55 and 15; 5: 84.
32. Mandeville, *The Fable of the Bees, or, Private Vices, Publick Benefits*, ed. F. B. Kaye (Oxford: Clarendon Press, 1924), 2: 284.
33. Pope, *Works*, eds. Elwin and Courthope, 10: 395–96.
34. Bernard Le Boivier de Fontenelle, *A Plurality of Worlds*, trans. John Glanvill (1688; rpt. London: Nonsuch Press, 1929), p. 17.
35. See, for example, A. Pannekoek, *A History of Astronomy* (London: George Allen & Unwin, 1961), p. 305.
36. See also F. C. Haber's "Darwinian Evolution in the Concept of Time," *The Study of Time I*, eds. J. T. Fraser, F. C. Haber and G. H. Müller (New York: Springer-Verlag, 1972); and his "The Cathedral Clock and the Cosmological Clock Metaphor," *The Study of Time II*, eds. Fraser and Lawrence.
37. Henry Power, *Experimental Philosophy* (London: John Martin and James Allestry, 1663), pp. 188–89.
38. Edward Taylor, *Jacob Behmen's Theosophick Philosophy Unfolded* (London: Tho. Salusbury, 1691), p. 396.
39. Henry Vaughan, "The Evening-watch" and "The Night," *Poetry and Selected Prose*, ed. L. C. Martin (London: Oxford University Press, 1963), pp. 256 and 359.
40. Henry Fielding, *Tom Jones*, in *Works*, ed. William Ernest Henley (London: William Heinemann, 1903), 5: 296–97 and 221.
41. Baldassare Castiglione, *The Book of the Courtier*, trans. Sir Thomas Hoby (London: J. M. Dent, 1928), pp. 309–10.
42. Donne, *Poems*, ed. Grierson, 1: 275–76. Grierson notes a parallel in Webster's *White Devil*, 1.2.313.
43. Ibid., 1: 246 and 129.
44. George Herbert, "Even-song," *Works*, ed. F. E. Hutchinson (Oxford: Clarendon Press, 1945), p. 64.
45. Vaughan, *Poetry and Selected Prose*, p. 256.
46. Robert Herrick, *Hesperides* (1648), in *Complete Poetry*, ed. J. Max Patrick (New York: New York University Press, 1963), p. 269; and John Milton, *Works*, ed. Frank Allen Patterson (New York: Columbia University Press, 1931–38), 4: 319.

47. Milton, *Works*, 1.1.33.

48. Ibid., 1.1.81. After writing this, I was delighted to discover that, though he has not noted the poem on Hobson, Claud Adelbert Thompson makes at length the case for the two-handed engine being a jack of the clock: "'That Two-Handed Engine' Will Smite: Time Will Have a Stop," *Studies in Philology* 59 (April 1962): 184–200.

49. Samuel Richardson, *Pamela* (London: J. M. Dent, 1914), 1: 339.

50. Dryden, *Works*, ed. Scott, 6: 194.

51. Sir Edward Bulwer-Lytton, *The New Timon*, 3rd ed. (London: Henry Colburn, 1846), 1.2.13.

52. George Gordon, Lord Byron, *Childe Harold's Pilgrimage*, in *Works*, ed. Thomas Moore (London: John Murray, 1832), 8: 238; 12: 251.

53. William Hazlitt, "On Cant and Hypocrisy," *Works*, ed. P. P. Howe (London and Toronto: J. M. Dent, 1933), 17: 350.

54. Oliver Wendell Holmes, *The Autocrat of the Breakfast Table*, in *Works* (Boston and New York: Houghton Mifflin, 1891), 1: 185–87, and 191–92. *The Autocrat* was first copyrighted in 1858.

55. John Pomfret, *Poems Upon Several Occasions* (London: J. Walthoe, 1731), p. 5.

56. Walter Savage Landor, "On Man," *Poetical Works*, ed. Stephen Wheeler (Oxford: Clarendon Press, 1937), 3: 233.

57. Archdeacon Paley's *Natural Theology* (1802) went into twenty editions by 1820, and in 1885 was "Revized to harmonize with modern science."

58. For several of the references to Dickens, I am indebted to Frederick R. S. Rogers, "Charles Dickens and Horology," *Antiquarian Horology* 7 (December 1970): 60–65.

The Poets' Reaction to Cartesian Clockwork

1. C. S. Duncan, "The Scientist as a Comic Type," *Modern Philology* 14 (1916): 281–91.

2. Jonathan Swift, *The Prose Works*, ed. Temple Scott (London: George Bell, 1898), 4: 278. This is not included in *Works*, ed. Davis.

3. Jonathan Swift, *Works*, ed. Herbert Davis (Oxford: Basil Blackwell, 1965), 2: 155.
4. William Wycherley, *Complete Plays*, ed. Gerald Weales (New York: University Press, 1967), p. 261.
5. Laurence Sterne, *The Life and Opinions of Tristram Shandy, Gentleman*, ed. Samuel Holt Monk (New York: Holt, Rinehart, 1950), p. 158.
6. Jonathan Swift, *A Tale of a Tub to Which Is Added the Battle of the Books and the Mechanical Operation of the Spirit*, ed. A. C. Guthkelch and D. Nichol Smith, 2nd ed. (Oxford: Clarendon Press, 1958), pp. 289, 275–76, and 192–94 (references to the "*Tale of a Tub* trilogy" will be made to this edition); and Swift, *Works*, ed. Davis, 11: 159–60.
7. Pope, *Poems* (Twickenham) ed. E. Audra and Aubrey Williams (London: Methuen, 1961), 5: 386–87.
8. Matthew Prior, *The Poetical Works*, ed. Reginald Brimley Johnson (London: George Bell, 1892), 2: 71.
9. Arthur Hugh Clough, *The Poems and Prose Remains*, edited by his wife (London: Macmillan, 1869), 2: 84–85.
10. Berkeley, *Works*, ed. A. A. Luce and T. E. Jessop (London: Thomas Nelson, 1948–56), 7: 185–86.
11. Swift, *Works*, ed. Davis, 11: 35.
12. G. H. Baillie and C. A. Ilbert, *Britten's Old Clocks and Watches and Their Makers*, 7th ed. (New York: Bonanza Books, 1956), p. 256.
13. Wilbur L. Cross, *The Life and Times of Laurence Sterne*, 3rd ed. (New Haven: Yale University Press, 1929), pp. 228–29.
14. Anon., *The Clockmakers Outcry against the Author of the Life and Opinions of Tristram Shandy* (London: J. Burd, 1760), pp. viii, 13–15, and 19. Mr. J. C. T. Oates—author of the "Shandeana" section of the revised *C.B.E.L.*—has kindly permitted me to examine his personal copies of the first and fourth editions of this rare pamphlet. He is not opposed to the possibility that it may have been written by Sterne himself.
15. Ibid., pp. 38–44.
16. Boileau, *Oeuvres complètes*, ed. Françoise Escal (Paris: Bibliothèque de la Pléiade, 1966), p. 1006.
17. Samuel Johnson, *The Idler* and *The Adventurer*, vol. 2 in *Works*, ed. W. J. Bate and others (New Haven and London: Yale University Press, 1963), pp. 82–83.
18. See the useful article: Frank Brady, "*Tristram Shandy*: Sexuality, Morality, and Sensibility," *Eighteenth Century Studies* 4 (Fall 1970): 41–56.
19. Like Hobbes, Walter confuses the "body national" with the "body natural" (1.18). Swift seems to mock Hobbes for this in *Gulliver* 3.6; Plato uses the comparison seriously in *Republic* bk. 4.
20. 1.4, 3.18, *passim*; see Locke, *Essay*, 2.33.
21. Bernard Caillard, "The History of the Pendulum Watch,"

Antiquarian Horology 3 (March 1960): 41–43.

22. Pope, *Works*, ed. Whitwell Elwin and William John Courthope, (London: John Murray, 1886), 10: 335.

23. Prior, *Works*, 2: 69–70.

24. Robert Browning, *The Complete Works*, ed. Charlotte Porter and Helen A. Clarke (New York: Riverdale Press, 1903), 5: 197.

25. Leonora Cohen Rosenfield, *From Beast-Machine to Man-Machine*, enlarged ed. (New York: Octagon Books, 1968), p. 183.

26. Fontaine, *Mémoires pour servir à l'histoire de Port-Royal* (Cologne, 1738), 2: 52–53 and 470–71.

27. Antoine Houdar de la Motte, *One Hundred New Court Fables*, trans. Samber (London: E. Curll, 1721), p. 36.

28. Thomas Brown, *Works*, ed. James Drake, 7th ed. (London: Edward Midwinter, 1730), 1: 278–80; Fontenelle, *Lettres Galantes*, ed. Daniel Delafarge (Paris: Société d'Edition "Les Belles Lettres," 1961), pp. 53–54. If "winding up the clock" had acquired the popular meaning here implied, it would reinforce the point made in the *Clockmakers Outcry* and add piquancy to Walter's role in Sterne's *Tristram*.

29. Bougeant, *A Philosophical Amusement upon the Language of Beasts* (London: T. Cooper, 1739), pp. 3–4.

30. Walter Shugg, "The Cartesian Beast-Machine in English Literature," *Journal of the History of Ideas* 29 (1968): 202; John Locke, *Some Thoughts concerning Education* (1693; rpt. Menston: Scolar Press, 1970), p. 230. See also Pierre Garai, "Le Cartésianisme et le classicisme anglais," *Revue de Littérature Comparée* 31 (1957): 374–75 and 378.

31. Allardyce Nicoll, *Restoration Drama*, 4th ed., vol. 1 of *A History of English Drama 1660–1900* (Cambridge: University Press, 1967), pp. 263–67.

32. James Boswell, *Life of Johnson*, ed. George Birkbeck Hill and L. F. Powell (Oxford: Clarendon Press, 1934–50), 2: 54.

33. [Daniel Defoe, ed.] *The Review* (1705; rpt. New York: Columbia University Press, 1938), 2: 15.

34. Ibid., 2: 26.

35. Swift, *Works*, ed. Davis, 11: 242.

36. Ibid., 11: 272–73, 290, and 295.

37. John Ray, *The Wisdom of God Manifested in the Works of the Creation*, 9th ed. (London: William and John Innys, 1727), pp. 54–55.

38. Bernard Mandeville, *The Fable of the Bees, or, Private Vices, Publick Benefits*, ed. F. B. Kaye (Oxford: Clarenden Press, 1924), 2: 285–88.

39. Sir Richard Blackmore, *Essays upon Several Subjects*, 2nd ed. (London: G. Grierson, 1716), pp. 36–37.

40. Robert Boyle, *Works* (London: W. Johnston and others, 1772), 2: 71.

41. Abraham Cowley, *Prose Works* (London: William Pickering, 1826), p. 214.

42. Samuel Johnson, *The Rambler*, vol. 1 in *Works*, ed. Bate, pp. 222–23.
43. Prior, *Works*, 1: 124–25.
44. William Somerville, *The Poetical Works* (London: C. Cooke, 1801), p. 96.
45. Swift, *Works*, ed. Davis, 11: 35.
46. Anon., 3rd ed. (London: Andrew Crook, 1665), p. 397.
47. Cited from Derek J. de Solla Price, "On the Origin of Clockwork, Perpetual Motion Devices, and the Compass," *United States National Museum Bulletion 218* (Washington, D.C.: Smithsonian Institution, 1959), p. 90 (Yonge's translation).
48. Bolingbroke, *Works* (Philadelphia: Carey and Hart, 1841), 3: 281.

Augustan Clockwork and Romantic Organicism

1. Émile Krantz, *Essai sur l'esthétique de Descartes: Rapports de la doctrine cartésienne avec la littérature classique française au XVIIe siècle* (1898; rpt. Geneva: Slatkine Reprints, 1970); Marjorie Nicolson, "The Early Stage of Cartesianism in England," *Studies in Philology* 26 (1929): 356–74; and Emerson R. Marks, *The Poetics of Reason* (New York: Random House, 1968), pp. 4 and 45.
2. I use the term *neo-classical* in its literary sense as being synonymous with English Augustan literature. Though Donald Greene and others have argued cogently that such terms as *neo-classicism* and *romanticism* are over-simplifications, I feel obliged to use them, albeit circumspectly, as part of our critical heritage. See *The Age of Exuberance* (New York: Random House, 1970), p. 159f.
3. William Cowper, *Poetical Works*, ed. H. S. Milford, 4th ed. (London: Oxford University Press, 1967), pp. 12–14.
4. A. W. Schlegel, *Vorlesungen über Schöne-Litteratur und Kunst*, in *Deutsche Litteraturdenkmale des 18. und 19. Jahrhunderts* (Heilbronn: Gebr. Henninger, 1884), pp. 7–8 and 102.
5. Ludwig Tieck, *Schriften* (Berlin: G. Reimer, 1828), 1: 15.
6. M. H. Abrams, *The Mirror and the Lamp* (New York: W. W. Norton, 1958), p. 186. The chapter on "Mechanical and Organic

Theories" is particularly helpful. Sussman demonstrates that Wells and even Kipling eventually succumbed to the "antimachine modes of the nineteenth century" (*Victorians and the Machine*, pp. 195–96, 193, and 220), as had Carlyle, Dickens, Ruskin, Morris, and Butler before them. Though heavily influenced by technology all but Wells are shown to have a varying measure of organic, vitalist, or pastoral orientation in their writings (pp. 14, 79, 104, 138, 151, 197, 200, 226, and 232).

7. Abrams, *Mirror*, p. 58.

8. William Powell Jones, *The Rhetoric of Science* (Berkeley and Los Angeles: University of California Press, 1966), p. 17. Jones points out that "After 1760 the most popular subject for scientific study was natural history, especially botany."

9. Edward Yong, *Conjectures on Original Composition*, ed. Edith J. Morley (Manchester: University Press, 1918), pp. 19 and 7.

10. Abrams, *Mirror*, pp. 68–69.

11. Samuel Taylor Coleridge, *Biographia Literaria*, ed. George Watson (London: Dent, 1965), p. 60.

12. Ibid., pp. 91–92 and 167.

13. M. H. Abrams, "The Correspondent Breeze: A Romantic Metaphor," in *English Romantic Poets*, ed. M. H. Abrams (New York: Oxford University Press, 1960), p. 38.

14. Samuel Taylor Coleridge, *Complete Poetical Works*, ed. Ernest Hartley Coleridge (Oxford: Clarendon Press, 1912), 2: 1022–23; 1: 102.

15. René Descartes, *L'Homme*, and *Un Traité de la formation du foetus* (Paris: Charles Angot, 1664), p. 112.

16. Cowper, *Works*, pp. 255 and 105.

17. Ibid., p. 411.

18. Walter Thornbury, *The Life of J. M. W. Turner* (London: Hurst and Blackett, 1862), p. 16.

19. John Dryden, *Works*, ed. Walter Scott (London: James Ballantyne, 1808), 3: 340.

20. Pope, *Poems* (Twickenham), ed. E. Audra and Aubrey Williams (London: Methuen, 1961), 1: 249.

21. Ibid., 5: 357.

22. William Hogarth, *The Analysis of Beauty* (London: printed for the author, 1753), p. 71.

23. Marjorie Nicolson, "Early Stage of Cartesianism," pp. 370–71.

24. Ibid., p. 373.

25. Ibid., p. 372.

26. Swift, *Works*, ed. Herbert Davis (Oxford: Basil Blackwell, 1965), 11: 197.

27. John Donne, *The Poems*, ed. Herbert J. C. Grierson (London: Oxford University Press, 1912), 1: 229 and 237–38.

28. Ibid., 1: 246.

29. R. F. Jones, *Ancients and Moderns*, 2nd ed. (Berkeley: University of California Press, 1961), p. 40.

30. Descartes, *L'Homme* (Paris: Charles Angot, 1664), pp. 106–107. See also John Morris, "Pattern Recognition in Descartes' Automata," *Isis* (Winter 1969): 451–55.
31. Dryden, *Works*, ed. Scott, 11: 214–16.
32. Aquinas, *Summa Theologica*, trans. Thomas Gilby (London: Eyre and Spottiswoode, 1970), 17: 129.
33. Thomas Hobbes, *The English Works*, ed. Sir William Molesworth (1840; rpt. n.p.: Scientia Aalen, 1962), 4: 448–49.
34. Thomas Rymer, *The Critical Works*, ed. Curt A. Zimansky (New Haven: Yale University Press, 1956), p. 18.
35. [Henry Barker], *The Polite Gentleman* (London: John Nutt, 1700), pp. 89–90 and 93.
36. Alexander Gerard, *An Essay on Genius* (London: W. Strahan, 1774), p. 49.
37. Ibid., pp. 265 and 50.
38. Coleridge, *Biographia*, pp. 85–86.

The Ambivalence of the Novelists

1. James Boswell, *London Journal 1762–1763*, ed. Frederick A. Pottle (New York: McGraw-Hill, 1950), pp. 265–66, *passim.*
2. Johann Wolfgang von Goethe, *Wilhelm Meister's Apprenticeship and Travels*, trans. Thomas Carlyle, ed. Clement King Shorter (Chicago: A. C. McClurg, 1890), 2: 44. (This passage is from "Appendix B" translated by Edward Bell.)
3. G. H. Baillie and C. A. Ilbert, *Britten's Old Clocks and Watches and Their Makers*, 7th ed. (New York: Bonanza Books, 1956), p. 458; and Sir Walter Scott, *The Fortunes of Nigel* (London: Dent, 1969), pp. 437–38.
4. Scott, *Nigel*, pp. 25 and 427.
5. Lewis Carroll [Charles Lutwidge Dodgson], *Alice in Wonderland* [and *Alice Through the Looking Glass*], ed. Donald J. Gray (New York: W. W. Norton, 1971), p. 282. With regard to Carroll's "mechanical" habits and predilection for order and regularity, see pp. 306, 252, 291, 314, and 405.
6. Ibid., pp. 55–56.

7. Cited from Frederick R. S. Rogers, "Charles Dickens and Horology," *Antiquarian Horology* 7 (December 1970): 64.

8. Charles Dickens, *Master Humphrey's Clock* and *A Child's History of England* (London: Oxford University Press, 1958), pp. 84 and 106–108. It is worth noting that Big Ben is never mentioned, though Grimthorpe's famous clock was working at least ten years before Dickens died.

9. Ibid., pp. 93–94.

10. "Coventry's Watch," *The Dickensian* 47 (June 1951): 17; and Rogers, "Charles Dickens," p. 63.

11. Charles Dickens, *Great Expectations* (New York: Holt Rinehart, 1948), p. 328.

12. Thomas Hardy, *Far from the Madding Crowd* (London: Macmillan, The New Wessex Edition, 1974), p. 132.

13. Mary Hamer, "Working Diary for *The Last Chronicle of Barset,*" *Times Literary Supplement* (24 December 1971), p. 1606.

14. Anthony Trollope, *An Autobiography* (London: Oxford University Press, 1950), pp. 118–19 and 271–72.

15. Cited from Rogers, "Charles Dickens," p. 62.

16. Significantly Sussmann—though his study is not concerned with clocks—relates diabolism to machinery in each of seven chapters on typical Victorians. See *Victorians and the Machine* pp. 25, 50, 100–103, 127, 144–46, 171–74, 181–82, and 224.

17. Walt Whitman, *Leaves of Grass*, ed. Harold W. Blodgett and Sculley Bradley (New York: Norton, 1965), pp. 411–12.

18. Leo Marx, *The Machine in the Garden: Technology and the Pastoral Ideal in America* (New York: Oxford University Press, 1964), pp. 15–16. Though his work deals primarily with the impact of steam technology, I am indebted to Mr. Marx for a number of my observations on the American novelists.

19. Nathaniel Hawthorne, *The American Notebooks*, ed. Randall Stewart (New Haven: Yale University Press, 1932), p. 104.

20. Henry D. Thoreau, *Walden*, ed. Walter Harding (New York: Twayne Publishers, Inc., 1962), p. 108.

21. Ibid., pp. 49, 27, 89, 88, 106, 64, 114, and 224.

22. According to Leo Marx, Thoreau must also depend on the imagination for his vision of the melting railroad bank, Walden's "climactic trope: a visual image that figures in the realization of a pastoral ideal in the age of the machines" (*Machine in the Garden*, p. 261).

23. Thoreau, *Walden*, p. 110.

24. Herman Melville, *The Romances* (New York: Tudor Publishing Co., 1931), pp. 1088–89, 1090, and 1086.

25. Ibid., pp. 859–60

26. Quoted in Marx, *Machine in the Garden*, p. 278.

27. Herman Melville, *Pierre, or the Ambiguities*, ed. Harrison

Hayford, Herschel Parker, and G. Thomas Tanselle, *The Writings of Herman Melville*, vol. 7 (Evanston, Illinois: Northwestern University Press, 1971), pp. 211 and 212.

The Clockwork Devil

1. A. A. O. Fox ed. *An Anthology of Clocks and Watches* (Swansea: published by the editor, n.d.), pp. 10–11. For this reference and for other generous advice on horological history, I am indebted to Mr. Beresford Hutchinson of the Department of Medieval and Later Antiquities at the British Museum.
2. Rabelais, *Works*, 1: 266 (1.53).
3. Cited from Lawrence Wright, *Clockwork Man* (London: Elek Books, 1968), p. 29.
4. George Farquhar, *Complete Works*, ed. Charles Stonehill (London: Nonesuch Press, 1930), 1: 293 (1.1); Locke, *Education*, pp. 14–15.
5. William Blake, *The Poetry and Prose*, ed. David V. Erdman (New York: Doubleday, 1970), p. 214.
6. Charles Lamb, *Essays of Elia* (London: Dent, 1962), pp. 97–98.
7. Ben Jonson, *Works*, ed. W. Gifford (London: Bickers and Son, 1875), 5: 393 (4.3).
8. Carroll, *Alice*, pp. 56–57.
9. Charles Baudelaire, "L'Horloge," *Les Fleurs du mal*, ed. Antoine Adam (Paris: Éditions Garnier Frères, 1959), pp. 87 and 374.
10. Alfred Tennyson, *The Devil and the Lady*, ed. Charles Tennyson (London: Macmillan, 1930), p. 21 (1.5).
11. Alfred Chapuis and Edmond Droz, *Automata*, trans. Alec Reid (Neuchatel: Griffon, 1958); A. G. Drachman, *The Mechanical Technology of Greek and Roman Antiquity* (Madison: University of Wisconsin Press, 1963).
12. Both papers appeared in *Technology and Culture* 5 (1964): 9–42.
13. Vaucanson, *An Account of the Mechanism of an Automaton*..., trans. J. T. Desaguliers (London: Stephen Varillon, 1742), title page.
14. (New York: W. W. Norton, 1965), pp. 64–66.
15. Diane Perrot "*The Grotto* the Long Lost Automata by Jaquet

Droz," *Antiquarian Horology* 5 (December 1966): 170–72; and Bedini, "Role of Automata," p. 39.

16. Price, "Automata," pp. 11–12; see also Merriam Sherwood, "Magi and Mechanics in Medieval Fiction," *Studies in Philology* 44 (October 1947): 567–92.

17. Clemens Brentano, *Gockel, Hinkel und Gackeleia* in *Die Deutschen Romantiker*, ed. Gerhard Stenzel (Salzburg: Verlag "Das Bergland-Buch," n.d.), 2: 604, 614, 620, *passim.*

18. Fernand Baldensperger, "Une Nouvelle Française peu connue sur le machinisme menaçant," *Modern Language Notes* (May 1945): 321–23.

19. E. T. A. Hoffmann, *Selected Writings*, ed. and trans. Leonard J. Kent and Elizabeth C. Knight (Chicago and London: University of Chicago Press, 1969), 1: 163.

20. Andreas Gryphius, *Cardenio und Celinde* in *Zwei Trauerspiele*, ed. Erik Lunding (Copenhagen: Gyldendals Forlag, 1938), p. 41 (l.1354).

21. Edgar Allan Poe, *The Complete Tales and Poems* (New York: Modern Library, 1938), pp. 115 and 105.

22. Tieck, *Schriften*, 1: 15–16.

23. Samuel Butler, *The Works* (Shrewsbury edition), ed. Henry Festing Jones and A. T. Bartholemew (London: Jonathan Cape, 1923), 1: 236–37; machines like animals tend to become smaller as they evolve, "take the watch for instance.... this little creature is but a development of the cumbrous clocks of the thirteenth century" (*Works*, 1: 210).

24. Aldous Huxley, *Brave New World* (Harmondsworth: Penguin, 1965), pp. 27 and 18. The Expressionists, after World War I, were fascinated with the impact of machinery on man.

25. Samuel Bulter, *Erewhon or Over the Range* (London: Jonathan Cape, 1921), pp. 63–64.

Index

Device of Christophe Plantin

This book has been set in Plantin by Asco Trade Typesetting Ltd. of Hong Kong. Christophe Plantin was French born but emigrated because of religious persecution to Belgium where he began printing and publishing books in 1555. He was the leading printer of the second half of the sixteenth century.

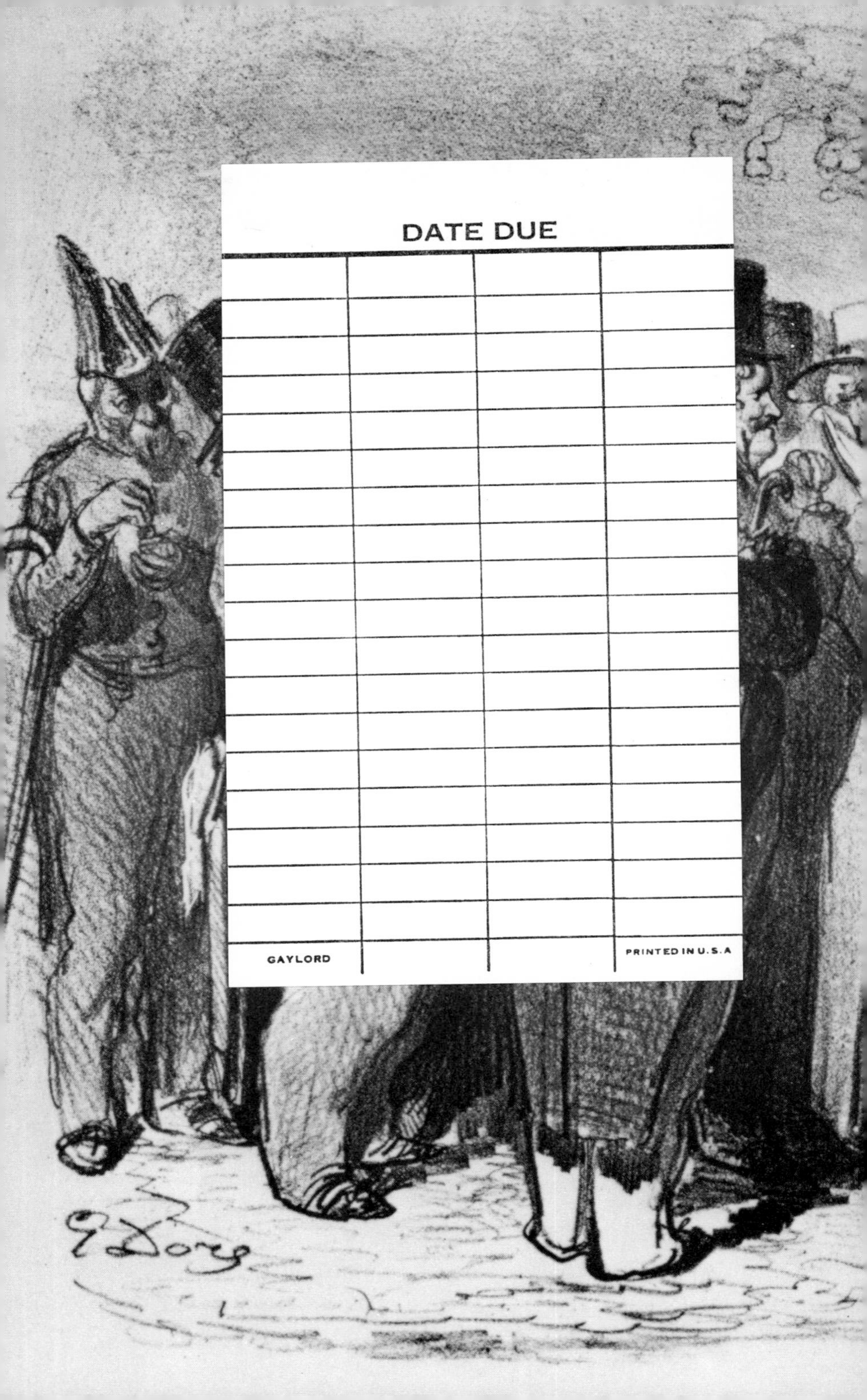
DATE DUE
GAYLORD
PRINTED IN U.S.A.
G. Doré